Living with the Sea

The seas and oceans are currently taking centre stage in academic study and public consciousness. From the plastics littering our seas and the role of climate change on ocean currents, to unequal access of marine resources and the treacherous experiences of seafarers who keep our global economy afloat, now is a crucial time to examine how we live with the sea.

Although an 'oceanic turn' is taking place with a burgeoning of academic work which takes seriously the place of seas and oceans in understanding socio-cultural and political life, past and present, there is a significant gap concerning the ways in which we engage with seas and oceans with a will to enliven action and evoke change.

This ambitious book brings together an interdisciplinary and international cohort of contributors from within and beyond academia offering a range and diversity of chapters, unlike previous collections. This book will appeal to a wide audience – from spatial planning, architectural design and geography, to educational studies, anthropology and cultural studies. An examination through these lenses can help us to better understand human relationships with the seas and oceans and promote an ethic of care for the future.

Mike Brown is the General Manager of Coastguard Boating Education, New Zealand. He holds an honorary Research Fellowship at Auckland University of Technology. Published works include *Adventurous Learning: A Pedagogy for a Changing World* (2016, with S. Beames), and *Seascapes: Shaped by the Sea* (2015, with B. Humberstone). Mike has been involved in sail training and has cruised the South West Pacific by sailboat. He lives on his yacht within a few hundred metres of his workplace.

Kimberley Peters teaches Human Geography at the University of Liverpool. Her research seeks to better understand the governance of maritime and other 'non-grounded' spaces. She is a member of Liverpool's Institute for Sustainable Coasts and Oceans (LISCO) and the Centre for Port and Maritime History. Her work includes *Water Worlds: Human Geographies of the Ocean* (2014, with J. Anderson), *The Mobilities of Ships* (2015, with A. Anim-Addo and W. Hasty) and *Carceral Mobilities* (2017, with J. Turner). Kimberley has written over 30 peer-reviewed articles and book chapters.

Routledge Studies in Human Geography

This series provides a forum for innovative, vibrant, and critical debate within Human Geography. Titles will reflect the wealth of research which is taking place in this diverse and ever-expanding field. Contributions will be drawn from the main sub-disciplines and from innovative areas of work which have no particular sub-disciplinary allegiances.

Towards a Political Economy of Resource-dependent Regions
Greg Halseth and Laura Ryser

Crisis Spaces
Structures, Struggles and Solidarity in Southern Europe
Costis Hadjimichalis

Branding the Nation, the Place, the Product
Edited by Ulrich Ermann and Klaus-Jürgen Hermanik

Geographical Gerontology
Edited by Mark Skinner, Gavin Andrews, and Malcolm Cutchin

New Geographies of the Globalized World
Edited by Marcin Wojciech Solarz

Creative Placemaking
Research, Theory and Practice
Edited by Cara Courage and Anita McKeown

Living with the Sea
Knowledge, Awareness and Action
Edited by Mike Brown and Kimberley Peters

For more information about this series, please visit: www.routledge.com/Routledge-Studies-in-Human-Geography/book-series/SE0514

Living with the Sea

Knowledge, Awareness and Action

Edited by Mike Brown and Kimberley Peters

Routledge
Taylor & Francis Group
LONDON AND NEW YORK

First published 2019
by Routledge
2 Park Square, Milton Park, Abingdon, Oxon OX14 4RN

and by Routledge
52 Vanderbilt Avenue, New York, NY 10017, USA

First issued in paperback 2020

Routledge is an imprint of the Taylor & Francis Group, an informa business

British Library Cataloguing-in-Publication Data
A catalogue record for this book is available from the British Library

Library of Congress Cataloging-in-Publication Data
A catalog record has been requested for this book

ISBN 13: 978-0-367-58692-8 (pbk)
ISBN 13: 978-1-138-06207-8 (hbk)

Typeset in Times New Roman
by Integra Software Services Pvt. Ltd.

Contents

Figures and tables

Figures

Tables

Contributors

Mick Abbott is Associate Professor at Lincoln University's School of Landscape Architecture. His research within the university's DesignLab investigates how social, cultural and economic value can be built out of strategies that increase biodiversity. Current design-directed research projects include the Eden Project NZ, Ngai Tahu Farming's 7600ha Eyrewell dairy development project, and Rio Tinto's Paparoa National Park mining restoration project. Edited books on landscape themes include *Wild Heart: The Possibility of Wilderness in Aotearoa New Zealand* (2011), *Making Our Place: Exploring Land-use Tension in Aotearoa New Zealand* (2011), and *Beyond the Scene: Perspectives on Landscape in Aotearoa New Zealand* (2010).

Maria Borovnik is a Senior Lecturer in Development Studies at Massey University, New Zealand. Her research focuses on the intersections between mobility, migration and development with a specific angle on mobile livelihoods of Pacific seafarers and their families. Her publications have focused on transnational aspects of labour circulation, identity, health and well-being of seafarers, and on how the actual experience of work on cargo ships influences perceptions of self, ship, port and home communities. She serves as co-editor for the Working Paper Series of the New Zealand Mobilities Network, and as book review editor for the *New Zealand Geographer*.

Jacky Bowring is Professor of Landscape Architecture at Lincoln University. Jacky's research and teaching interests circle areas of cultural landscape, history, memory and emotion. She is editor of the peer-reviewed journal *Landscape Review*, and author of *A Field Guide to Melancholy* (2008) and *Melancholy and the Landscape: Locating Sadness, Memory and Reflection in the Landscape* (2017). Recent publications reflect the response to Christchurch's earthquakes, particularly in the areas of 'design as caring', dwelling, memory and melancholy. As a landscape architect she also explores ideas through designing and in 2017 was one of five winners in journal *LA+ 's* global competition to design an island, for the 'Island of Lost Objects'.

Lars Brabyn is a Senior Lecturer in the Geography and Environmental Planning programmes at the University of Waikato. He principally teaches Geographical

Information System but has an interest in landscapes and associated values. As part of his PhD and in subsequent years, he developed the New Zealand Landscape Character Classification.

Easkey Britton is a Post-Doctoral Research Fellow in Social Innovation at the Whitaker Institute, National University of Ireland. She co-leads the interdisciplinary *NEAR-Health* project investigating the use of blue space to restore health and wellbeing. She channels her passion for surfing and the sea into social change. Her parents taught her to surf at age four and her life has revolved around the sea and surfing. Her work is influenced by the ocean and the lessons learned pioneering women's big-wave surfing led to a TEDx talk in 2013: *Just Add Surf.* Passionate about creative and collaborative processes, she founded *Like Water*, a platform to explore innovative ways to reconnect with who we are, our environment and each other, through water.

Chris Eames is a Senior Lecturer in Environmental Education (EE) at the University of Waikato in Hamilton, New Zealand. He works with pre-service teachers, and with many postgraduate students, with a particular focus on education practice. He also advocates at national level and works at a local level to promote EE and to protect and restore the natural environments in our country. In his spare time he enjoys being in those environments with family and friends, who share his sense of connection to these places.

Barbara Humberstone is a Professor of Sociology of Sport and Outdoor Education at Buckinghamshire New University, UK. Her research interests include: embodiment, alternative physical activities and life-long learning; embodiment and water/ seascapes; and wellbeing and outdoor pedagogies. She co-edited *Seascapes: Shaped by the Sea: Embodied Narratives and Fluid Geographies* (2015) and *International Handbook of Outdoor Studies* (2016). She is co-editor of the new Routledge series *Advances in Outdoor Studies* and has published papers in a variety of journals. She is editor of *Journal of Adventure Education and Outdoor Learning* and was Chair of the European Institute for Outdoor Adventure Education and Experiential Learning. She is a keen windsurfer, walker, swimmer and yogini.

David Irwin is the programme manager of Sustainability and Outdoor Education at Ara Institute of Canterbury, located in Christchurch, Aotearoa New Zealand. His research and teaching interests lie in the exploration of culture, identity and human–nature relationships, education for sustainability and social change in an organisational context. Dave is editor of *Te Whakatika*, a bi-annual journal for education outside the classroom, and co-editor of *Outdoor Education in Aotearoa New Zealand: A New Vision for the Twenty First Century* (2012). Dave has lived beside, in and on the ocean for most of his life.

Robin Kearns is Professor of Geography in the University of Auckland's School of Environment. His research has traversed various domains in social and cultural geography including housing, health, neighbourhoods and rural communities. A unifying concern in his publications has been developing links between the nature of place and human wellbeing. He is an editor of the journal *Health & Place*. His book *The Afterlives of the Psychiatric Asylum* has seen a return to the mental health care theme of his PhD. Other recent books have explored children's experience of health in cities and the place of popular music in human wellbeing. Bringing a research perspective to a lifelong love of the coast and islands is a current preoccupation.

Mark Leather is a Senior Lecturer and the Programme Lead for Adventure Education and Outdoor Learning at the University of St Mark & St John in Plymouth, Devon, UK. He specialises in outdoor and experiential learning whilst teaching on undergraduate and postgraduate education programmes that utilise the outdoors and natural environments. He enjoys thinking about his professional practice and personal leisure time, and researching and writing in this area. Mark enjoys connecting with people, places and the planet, especially on or near the sea: sailing, kayaking, paddle boarding, walking the coastal path or just sitting on the beach, silently staring at the horizon, whilst listening, smelling and feeling this special place.

lisahunter has played with and worked in relation to the sea for most of her life, fascinated with the subject positions offered, the practices encountered, and the sensorial experiences that emerge. Some of these were explored in recent publications such as 'Seaspaces: Surfing the sea as pedagogy of self' (2015); 'The long and short of (performance) surfing: Tightening patriarchal threads in boardshorts and bikinis' (2017); 'Desexing surfing? (Queer) Pedagogies of possibility' (2017); 'Positioning participation in the field of surfing: sex, equity, and illusion' (2018) and the book *Surfing, Sexes, Genders & Sexualities* (2018). lisahunter is the Founder and Director of the Institute for Women Surfers (Australasia) and is currently working for Monash University, Australia.

Alistair Moore is a sailor and educator. Al has a tertiary qualification in outdoor education and professional maritime certificates of competence. He recently left the employment of the New Zealand Sailing Trust and is preparing his yacht for a voyage to the south-west Pacific to escape the New Zealand winter. He has no immediate plans to return to NZ and will see where the winds blow.

Rebecca Olive is a Lecturer in the School of Human Movement and Nutrition Sciences at The University of Queensland. Her work has focused on sex/gender, localism and environmental politics in surfing, as well as the use of social media as a research space and method. Her work takes a feminist cultural studies approach, and draws on ethnographic methods, which has led to her interest in research methods, ethics, praxis and pedagogies. She has published in journals including *Sport, Education and Society, Sustainability* and *International Journal of Cultural Studies*, and co-edited *Women in Action Sport Cultures: Power, Identity and Experience* (2016, with Thorpe). She continues to publish in surf media, and to produce her blog, *Making Friends with the Neighbours*.

Raewyn Peart is Policy Director for the Environmental Defence Society and has extensive experience in environmental law and policy. Raewyn's work has focused on landscape protection, coastal development and marine management. She has published *Castles in the Sand: What's Happening to the New Zealand Coast?* (2009), and *Dolphins of Aotearoa: Living with Dolphins in New Zealand* (2013). She co-authored a guide to marine management, *Sustainable Seas: Managing New Zealand's Marine Environment* (2015) and *The Story of the Hauraki Gulf* (2016). Raewyn was a member of the Stakeholder Working Group which completed a marine spatial plan for the Hauraki Gulf under the Sea Change – Tai Timu Tai Pari project. She is a keen sailor and snorkeller.

Susan Reid is a PhD candidate in the Department of Gender and Cultural Studies, University of Sydney. Drawing on environmental humanities, new materialist philosophies, law and contemporary art practices, her research explores dimensions of ethics and practices of mattering with living oceans. Susan has a multidisciplinary background as artist, curator, cultural developer, lawyer and environmental activist.

Nathaniel Trumbull is Director of the Maritime Studies Program of the University of Connecticut and Associate Professor of Geography. His research focuses on environmental and maritime topics related to the eastern Baltic Sea and Long Island Sound. His office is located in Mystic, Connecticut, on the by-the-sea Avery Point campus of the University of Connecticut. He has lived in Boston, New York City, Seattle and St. Petersburg (Russia). He is from Woods Hole, Massachusetts.

Nancy Vance is a Registered Landscape Architect and regular contributor as Lecturer at the School of Landscape Architecture, Lincoln University, New Zealand. With almost 20 years' experience consulting and teaching, and as many years sailing the shores of Banks Peninsula, her interests focus on reconceptualising the sea–land edge: the natural and cultural layers, the built interventions, their form and materials.

Belinda Wheaton is Associate Professor in Health Sport and Human Performance at the University of Waikato, NZ. She is best known for her research on the politics of identity in lifestyle sports, including the water-based activities surfing, windsurfing and kite-surfing, and also the politics of environmentalism. Publications include a monograph, *The Cultural Politics of Lifestyle Sports* (2013), and edited collections including *Understanding Lifestyle Sport* (2004) and *Leisure and the Politics of the Environment* (2013, with Mansfield). Although born and brought up in London (UK) she was introduced to sailing at an early age, and has spent most of her adult life living by the sea (swimming, sailing, windsurfing, body boarding, surfing, SUPing).

Acknowledgements

Our sincere thanks to the participants who saw merit in investing four days of their time in a conference with a difference. This book is evidence of your efforts to bring your expertise and experiences to a wider audience of the diverse ways in which we live with the sea.

Our thanks to Professor Bronwen Cowie, Director of the Wilf Malcolm Institute of Educational Research at the University of Waikato for supporting the conference financially. Further thanks go to the University of Liverpool for their support in enabling the transcription work and to Gabrielle Sale for the production of the index to help navigate this volume.

Our thanks to the team at Routledge for their patience as we continually extended the deadline on this collection. We appreciate your professionalism and understanding.

Mike would like to thank Kim for her perseverance and fortitude in the face of uncertainty – it was a big leap of faith to commit to coming sailing on the other side of the world. The final product would not have been as strong without your input. It was great to see you grasp the pleasure of being at sea on a small vessel. I hope it's the start of many enjoyable voyages.

Kim would like to thank Mike for bringing her on board to a new world of being *at* sea (rather than looking to it from the shore), and to this book project. Although I have made this process slower I am so grateful for the opportunity, and the time, to carefully work across so many disciplines, styles and approaches to academic work. This has greatly enriched my understanding of how we each live differently with the sea, and in turn each seek to make a difference through our engagements with this space.

To the team at the New Zealand Sailing Trust, thank you for being open to our requests and sharing your passion for making a difference in the world. Let us hope that in some small way this book also contributes to a greater understanding of the importance of experiences at sea as a way to a more sustainable future.

1 Introduction

Living with the sea: knowledge, awareness and action

Mike Brown and Kimberley Peters

In late 2017 the United Nations created a new resolution to tackle the growing issue of plastics and the pollution of the world's oceans. It would be one of the most significant, although not legally binding, acts of the twenty-first century for seeking to change how we live with the sea (UNEP, 2017). With the launch of the popular television programme *Blue Planet* also charting the devastating effects of our 'plastic planet' (Honeyborne and Brownlow, 2017), alongside international research highlighting the continued growth of the 'Great Pacific Garbage Patch' (Lebreton et al., 2018), the state of our seas and how we live with them has been brought into stark focus. In recent years we have witnessed other continued and increased strains and stresses on the 70 per cent of our world that is water. Research has attended to the detritus left by cruise liners (Lamers, Eijgelaar and Amelung, 2015); the invasive species that can emerge through emptying vessel ballast water (Holbech and Pedersen, 2018); the toxic waste of ship breaking (Demaria, 2016); and the carbon footprint of the container industry (Cidell, 2012). Beyond environmental harm alone, recent years have witnessed the desperate and horrifying realities of refugees travelling across treacherous seas (Coddington, 2018; Jones, 2016; Vives, 2017); the growing trade in sea space and sale of deep-sea mining exploration licences (Van Dover, 2011); the threat to vulnerable coastal communities through sea-level rises (McLean and Kench, 2015); the conflict over territory with the creation of artificial islands (Dolven et al., 2015). The list goes on. Now, more than ever, there is a growing acknowledgement that 'our world is an ocean world' (Langewiesche, 2001,1) but also that our ocean world is threatened – environmentally, socially, politically.

This acknowledgement has led to an 'oceanic turn' slowly emerging across the humanities and social sciences, with a burgeoning of academic work which takes seriously the place of seas and oceans in understanding socio-cultural and political life, past and present (see Mack, 2013; Ryan, 2012; Singer, 2014; Steinberg, 2001). This is alongside a continued development of ocean sciences: marine biology, oceanography and sea-based ecology. All this work is essential if we are to address concerns that have been raised about modern societies' lack of attention to the seas that surround and sustain us (Hau'ofa, 1998).

As editors of this collection we have contributed to this nascent oceanic project through the publication of edited works including *Waterworlds: Human Geographies of the Ocean* (Anderson and Peters, 2014) and

Seascapes: Shaped by the Sea (Brown and Humberstone, 2015). A key feature of the recent work in the humanities and social sciences has been to explore the seas and oceans as a 'fundamental space of human experience' (Steinberg, 2015, xii). Indeed, one of the central questions that has arisen is: how have, and do, humans engage with the more-than-human, mobile, encompassing, yet volatile space of the sea? (Braverman, 2015; Lehman, 2013; Peters, 2012; Steinberg, 2001). To date this question has been answered by drawing attention to the ways we engage with the sea through social constructions and representations of the oceans (in paintings, literature, maps and film, see Mack, 2013; Singer, 2014; Steinberg, 2001); through embodied practices and performances (from diving, to surfing, to sailing, see Brown and Humberstone, 2015); through work, trade and politics (see Anderson and Peters, 2014; Birtchnel, Savitzky and Urry, 2015; Cowen, 2014; Urry, 2014); and through policy and planning (Arico, 2015; Jay, 2008). What this existing body of work has achieved is a formidable and wide-ranging understanding *of* the seas and oceans, historically and in the contemporary climate, across a range of disciplines from history to planning, geography to cultural studies, anthropology to remote sensing, politics to the study of outdoor education.

Although highlighted above, it is beyond the scope of this introduction to review all of this literature in depth, though overviews are offered in individual chapters to follow. What is apparent, though, is a gap in this literature concerning the ways in which we engage *with* seas and oceans with a will to inspire action and evoke change. There have been excellent discussions of changes, frameworks and management techniques that attend to the human and physical plights that are shaping and impacting our oceans (see, for example, Arico, 2015). Such work, however, has lacked an embodied, personal account of how we interact with our ocean world and how these relations connect to our knowledge, awareness and action. This has been covered in some more popular literature (see Danson and D'Orso, 2011; Helvarg, 2010; Honeyborne and Brownlow, 2017), but it has yet to be taken on as a comprehensive, wide-ranging inter- and cross-disciplinary project for academics.[1] That is the gap this book seeks to fill.

An unusual project

The question of how we might write *with* and *for* the seas gave rise to a small conference hosted in the Hauraki Gulf, New Zealand, aboard a former round the world race yacht, owned by the New Zealand Sailing Trust. The genesis of this idea arose following the publication of *Seascapes* (Brown and Humberstone, 2015). Would it be possible to gather an international group of scholars with an interest in the sea and provide them with an experience of living together and being on the sea for a limited period? If so, what might be the contribution? The initial call for expressions of interest simply stated:

Seascapes 2016 provides a unique opportunity for scholars from a variety of disciplinary backgrounds to share research around the theme of living with the sea. The conference welcomes papers that will advance knowledge of our relationship with the sea and that have the potential for increasing awareness and action in the human and more-than-human environment ... Being prepared to live on board for 3 nights and be involved in communal living is an integral component of this conference.

(Email, Mike Brown to potential participants)

The 'venue' was the 80-foot yacht *Steinlager 2*, which became famous in the sailing world when the crew of the vessel won all six legs of the 1989/90 Whitbread Round the World Race under the command of the late Sir Peter Blake (Figure 1.1). Following his yacht racing career Sir Peter became involved in expeditions aimed at raising environmental awareness and he was appointed as a special envoy for the UN Environmental Programme (UNEP). It was while in the Amazon in late 2001 that his vessel was boarded and he was killed by armed robbers. *Steinlager 2* and Blake's previous boat *Lion New Zealand* are currently owned by the New Zealand Sailing Trust, whose inspiration comes from the adventures of Sir Peter, both as a leader of racing teams and as an environmental activist. The Trust exists to ensure that these key yachts are preserved and sailing voyages are conducted to help inspire the next generation to achieve their aspirations.

Figure 1.1 Steinlager 2, Kawau Island. Photo by Mike Brown.

The format of the 'floating' conference was a combination of sailing, social time, explorations ashore, and the presentation and discussion of papers – each an academically and personally situated contribution – that had been circulated prior to our meeting. At times throughout the trip we would gather in the cockpit of the vessel, or below decks if the weather was inclement, and listen and provide feedback on the draft chapters of what has now become this book. This proved to be a supportive and fruitful initiative. Given the intimate nature of the gathering, the intention was to foster a sense of community and provide the opportunity to discuss and refine ideas, rather than to 'deliver' a paper. As an outdoor educator who had previously been involved in sail training, Mike knew that the mere fact of being in a confined space, and being called upon to assist with sailing the boat and to help with the preparation and sharing of meals, would create a sense of community and familiarity that is difficult to cultivate in the 'traditional' conference format. People would often chat in pairs or small groups – sometimes this was related to the theme of the conference and at other times matters of a more general nature. Given the amount of laughter on board, there were certainly a few jokes or funny anecdotes being retold. When sailing, we were required to assist with hoisting the sails, steering the boat and trimming the sails to keep the boat moving efficiently. When anchored, people could go for a walk ashore, swim or relax on board with a 'cuppa'.

Emerging from a very particular setting, this book, then, is not a 'conventional' edited book. It is not a collation of chapters resulting from an established conference setting, or a dedicated call for contributions. It is a product of a time and a place, and an engagement of scholars who span a vast array of disciplines – science and social science – and who embrace qualitative and quantitative approaches. It is a result of the engagements of scholars who are at differing stages of their academic careers, from postgraduates to professors, as well as those who work beyond the academy, in policy, planning and architecture. At the heart of this book, and of the floating conference from which it developed, was a desire to engage in conversation. This book is a *dialogue* reflecting the varying conversations we had on the vessel about how we live with the sea; how we build knowledge of the sea, awareness of contemporary issues regarding this space and how we may bring about action for positive change.

As such, this book has not been easy to curate; nor should it have been. There are many different voices, styles and approaches. But to better understand how we live with the sea – to build knowledge, awareness and action – we arguably need to bring different academic subjects, policy approaches and personal perspectives together, blurring the boundaries between these so that we may come into conversation about the pressing issues related to our seas and oceans. Whilst there could be more engagement across the science–social science boundary in this project specifically, here we present the beginnings of such efforts. In order to attend to the issue of ocean plastics, for example, it is not just scientific evidence of ocean currents and ecosystem degradation that is needed, but also knowledge of how we may educate and inform future generations of how they might act in relation to purchasing, reusing and recycling plastics. This is a conversation that requires bringing together different scholars, with differing epistemological and ontological backgrounds.

Although this book is divided into discrete chapters, which each take on various approaches, advances, engagements and experiences with the sea, they sit together under the umbrella of this volume. Each chapter is not just a product of the author(s), but a product of *being with the sea, being with each other on the boat*. They are a result of cross-disciplinary debate and informal conversation. The book, in turn, has been organised to enable the reader to dip their toes into the water and into these varying disciplines, approaches and perspectives and to read them in situ. The book itself, in sum, does not sit under a singular discipline; it melds and blurs boundaries, much like the sea itself. Indeed, although the chapters may in themselves appear to sit in neat boxes, their contents often spill over, mixing literatures and methods. As such, whilst this is a book where chapters may be read independently by those interested in ocean heritage, planning or seafarers' working lives, each chapter also introduces a broader set of ideas and debates. The structure is also designed to encourage readers to move across chapters – to join our conversation – to take their knowledge, awareness and actions in relation to the sea into uncharted waters.

We were fortunate to attract an international gathering of attendees from a diverse range of disciplinary backgrounds to the conference, and resultantly to this book. That said, as the book is a product of a time, a place and an event, Aotearoa New Zealand is prominent in it, alongside chapters which take the reader to waters off the United States, Britain, Kiribati and Iran. The strong New Zealand focus of the collection is, however, important. On the one hand, as an island where many people live in such close proximity to the sea (see Chapters 3, 5 and 11), New Zealand offers an ideal thinking space for this project: of bringing people into touch with the sea to make sense of our connections to and responsibilities towards it. On the other hand, much literature on the seas and oceans that has formed part of the 'watery turn' in contemporary scholarship has focused stubbornly on the Global North. Work has long examined the Atlantic, North Sea and Mediterranean (see Armitage and Braddick, 2002; Braudel, 1966 [1995]; Gilroy, 1993; Peters, 2015; Rediker, 2007; Steinberg, 2001) at the expense of other seas and oceans. In spite of recent attention to Indian Ocean worlds (see, for example, Davies, 2013), the Global South remains neglected. This book therefore offers an important corrective.

The book that follows is a testament to the adventurous spirit and willingness of participants of the conference – the contributors to this book – to step outside the norms of academic work. It is this that has enabled us to bring to fruition such a wide-ranging text. What underpinned the success of this venture was the attendees' united passion for the well-being of our planet, based on a deeper understanding of people's relationship with the sea as a desire to bring about a change from our current trajectory. That is not to say the book is without its limits though. Of course, the economic means to travel, the physical ability to go offshore and the professional opportunities to work within settings which allow the time to engage in the pursuit of a boat-based conference is a privilege not all of those who live with the seas can enjoy. It is important to acknowledge that the conversations we had were partial and incomplete. It is also important to note that voices were absent. Whilst some chapters bring to the surface unheard stories and overlooked realities from our seas (see Chapters 8 and 9 especially), these are still framed and come *from* the scholars who articulate them. As such, this book might be taken as a starting point for a broader project of engaging an even wider set of disciplines, approaches, engagements and experiences – a wider set of individuals and embodied narratives – that help make sense of our relations with the sea. We hope, as editors, it may develop oceanic scholar Epeli Hau'ofa's call for collective action (1998), in order for all of us to retain our sense of humanity and community.

Outlining the book to come

In the collection that follows, our conversations are divided into two parts. In the first part we feature seven contributions under the heading 'Approaches and Advances'. This section takes us on a voyage of the various 'lenses' through which we may make sense of and engage more thoughtfully with the sea, and the novel methods we may use to capture and reflect upon our use of the seas. In Chapter 2, Bowring, Vance and Abbott approach the seas and oceans from the standpoint of landscape architecture. They continue the effort to challenge the stark binary between land and sea, demonstrating how the careful design of *seascape* architectures – of delineations, projections, overlays and sutures – can bring us into touch with the seas. Moving from architecture to planning, in Chapter 3 Peart examines the role of Marine Spatial Planning (MSP) for attending to the present and future use of our seas in ways that may better protect and conserve them. Offering a different perspective to much work in MSP, Peart doesn't present the outcomes of a marine spatial plan but rather reflects on the development of creating a plan, providing insights into this crucial process. Taking a different tack, in Chapter 4 Brabyn draws on quantitative methods and a classification system for assessing the values people assign to seascapes. Such an approach, he argues, is particularly important for policy-makers who rely on 'data' to rationalise action.

Shifting from architecture, planning and spatialised classifications, Chapter 5 considers how the disciplines of teaching, research and professional learning can evoke change through Eames' discussion of the 'action competence' approach. Engendering action competence, he contends, can foster sustainable behaviour in respect of our seas and oceans. Likewise, thinking through how we create better understandings of our marine and maritime world, in Chapter 6 Trumbull explores a heritage project with a difference: the restoration and sailing of the nineteenth century whaling vessel the *Charles W. Morgan*. Through the direct experience of the ship's 'Voyagers' and those who engaged from the shore, Trumbull demonstrates the possibilities of reflecting on the past to assist in making sense of the current state of our seas and oceans. Also focusing on self-reflection, lisahunter explores the potential of autoethnography and videographic methods for bringing us closer to the seas and oceans that we hope to understand and better care for. In Chapter 7, by piecing together experimental fragments of field texts, she alerts us to the need to find methods that allow us to consider our ethical sensibilities and to better communicate our knowledge, in order to build stronger responsibilities towards the seas. Finally, closing Part One, Reid is also concerned with our duty to our seas and oceans, the devastating impacts of anthropocentric change, and the shortcomings of governance regimes. Reid takes a materialist approach, centred on the physical character of ocean currents, knitting together scientific and social scientific knowledge. By giving the seas and oceans agency, she contends that we may understand the transitional nature of the ocean's dynamic systems and relations. In doing so she offers a conceptual grounding for finding better ways to live with and respond to challenging ocean futures.

Part Two of the book takes our 'conversations' in a different direction. Here we turn to consider the varying engagements and experiences that can provide oceanic knowledge, awareness and evoke environmental action. This section takes us to sea: to critically consider how we live with the sea for work and for leisure; how it is intimately tied to our own personal histories; and how these intimate knowledges may be the most powerful for building awareness and potential action. In Chapter 9, Borovnik draws on extensive offshore research with seafarers to give voice to their experiences of working with the sea. Arguably, no other group lives so viscerally with the conditions of our oceans. Borovnik draws our attention and awareness to the demanding realities of such employment – the tiring labour, the lack of sleep and the muddled temporal and spatial awareness of constant travelling. Such awareness is necessary to bring to light the experiences of those who keep our global world of flows moving. Chapter 10 also brings marginalised voices to the surface, questioning the often taken-for-granted assumptions of leisure engagements at sea. Here, Britton, Olive and Wheaton demonstrate that engaging with the sea is not a 'freedom' enjoyed by all. Coasts, shorelines and seas are contested spaces, unequally accessed and perpetually fought for. In bringing this knowledge to light, they demonstrate how individuals, carefully considered projects and communities themselves can bring about change. Also focusing on the capacity to bring about change, in Chapter 11 Irwin considers how we live with the knowledge that our oceans may be irreparably damaged. He explores how the next generation can come to terms with and still seek to protect our ocean world. Young people must understand the global and local issues around them; embrace behaviour change on their own terms; adapt to a changing world; and, ultimately, learn to value living in the moment.

The second half of Part Two turns from the voices of seafarers, surfers and students to personal, embodied experiences of being with the seas. The aim here is to consider how our own relations with the oceans forge knowledge, sensibilities, responsibilities and actions in respect of our oceans. In Chapter 12, Humberstone draws on her own experiences to explore how technology (a windsurfer) can dissolve the boundaries between us as humans and the more-than-human sea and its energies. Weaving together personal accounts with a theorisation of 'cyborgs', Humberstone argues that a return to myths, and to thinking of our relations with the sea beyond the planetary, engages us to think imaginatively and holistically about our oceanic ties, and of social and environmental justice and the ways in which we are merged with the planet and the cosmos. Following this, in Chapter 13, Leather reflects on his own past and how his positive associations with the seas have shaped his present knowledge and engagements with them, both personally and professionally. Leather warns of the romanticism of nostalgia in remembering our ties to the ocean; instead, we should be aware of how our memories are culturally situated. He urges us to critically use our own ocean pasts to consider our awareness and actions towards the oceans. Leading from this, Kearns also draws on his childhood experiences and engagements with the ocean but through less positive associations. Drawing from emotional geography and making sense of a fear of water, or 'bathophobia', Kearns offers an important reminder in this penultimate chapter that our relations with the sea can be deeply problematic; that we do not all share a passion for the sea, or, where we have strong feelings, they may not manifest in the same ways. This chapter acts as an essential point of departure from earlier chapters in considering the complexity of how we live with the sea. Whilst some of us have the pleasure of engaging with the sea, and feel compelled to take action for environmental 'good', for others the relationship may be underpinned by a sense of unease, of the danger of living and being with the sea. Kearns reminds us, then, that efforts to consider our watery engagements should be wide-ranging and all-encompassing. He sets in motion the trajectory we hope will continue from this book, where others may take up the project of thinking about how we live with the seas and how we form differing knowledges, build varied awareness and bring about a multitude of actions in a moment when the ocean is centre stage.

In Chapter 15, as we close this collection, we travel full circle back to this introduction, to the conference that inspired this book and to the *Steinlager 2*. We weave together a conversation with Alistair Moore, the skipper of the vessel, with our own insights and perspectives as editors. As will become apparent in this chapter, Alistair brings with him a wealth of sailing and educational experiences, yet this conference was a very different event compared to his usual voyages – typically with teenage school students. What became evident during the short period of time that we were together with Alastair on the boat was his interest in what was being discussed by our collective and his deep-seated knowledge of and love for the sea. In a typically understated Kiwi manner he had achieved a lot in the sailing world, and had in fact been a crew member on Sir Peter Blake's last voyage. We met with Alistair after the trip and sought to talk with him about his connection to the sea, bringing this together with our own relations to build some final thoughts and insights into the reality of the constantly shifting fluid world that is the sea.

Concluding thoughts

During the conference that preceded this book we came together in a compact community. We may have been physically confined within 80 feet of floating fibreglass, but we were able to share ideas freed from the disciplinary constraints and rigid time restrictions that are so much a part of large academic conferences. As will become evident in the following chapters, we brought with us very different experiences and levels of appreciation of 'being' at sea. For some, being on or in the sea was natural and commonplace, while for others the sea had been experienced less frequently or perhaps more abstractly. Yet for this short period of time we had the opportunity to overtly situate our writing and thinking 'within' a seascape. For a concentrated period our suspended fibreglass shell became our home. We moved about above – on deck – and below the watery surface, where the bunks were located under the waterline. We adjusted our movements in line with the motion of the boat – which was directly influenced by the sea, and the wind with it. We swam in buoyant waters. We inhaled the salt-laden air and felt the salt crystals dry on our skin. We were, at various times, in, on, ingesting and being coated by our liquid environment. This level of engagement and the ensuing contributions to *Living with the Sea* are intended to continue to foster an appreciation, love and call to action to act for and with the sea. For, as Hau'ofa has so eloquently stated:

> There are no more suitable people on earth to be the custodians of the oceans than those for whom the sea is home. We seem to have forgotten that we are such people. Our roots, our origins are embedded in the sea ... The sea is our pathway to each other and to everyone else, the sea is our endless saga, the sea is our most powerful metaphor, the ocean is in us.
>
> (Hau'ofa, 1998, 408)

Note

1 Large grant projects such as Ocean Governance for Sustainability, a pan-European ocean network initiative funded by the European Commission, are now moving in this direction and it is likely further books, special issue journal publications and other transdisciplinary projects will follow. For details of the project see this website: https://www.oceangov.eu/ (Accessed 22 March 2018).

References

Anderson, J. and Peters, K.(Eds.). (2014). *Water Worlds: Human Geographies of the Ocean.* Farnham: Ashgate.

Arico, S. (Ed.). (2015). *Ocean Sustainability in the 21st Century.* Cambridge: Cambridge University Press.

Armitage, D. and Braddick, M. J. (Eds.). (2002). *The British Atlantic World, 1500–1800.* Basingstoke: Sage.

Birtchnell, T., Savitzky, S. and Urry, J. (Eds.). (2015). *Cargo Mobilities: Moving Materials in a Global Age.* Abingdon: Routledge.

Braudel, F. (1966/1995). *The Mediterranean and the Mediterranean World in the Age of Philip II.* Berkeley, CA: University of California Press.

Braverman, I. (Ed.). (2015). *Animals, Biopolitics, Law: Lively Legalities*. Abingdon: Routledge.

Brown, M. and Humberstone, B. (Eds.). (2015). *Seascapes: Shaped by the Sea*. Farnham: Ashgate.

Cidell, J. (2012). Flows and pauses in the urban logistics landscape: The municipal regulation of shipping container mobilities. *Mobilities*, 7(2), 233–245.

Coddington, K. (2018). Settler colonial territorial imaginaries: Maritime mobilities and the 'tow-backs' of asylum seekers. In K. Peters, P. Steinberg and E. Stratford (Eds.). *Territory Beyond Terra*. London: Rowman and Littlefield.

Cowen, D. (2014). *The Deadly Life of Logistics: Mapping Violence in Global Trade*. Minneapolis, MN: University of Minnesota Press.

Danson, T. and D'Orso, M. (2011). *Oceana: Our Endangered Oceans and What We Can Do to Save Them*. Manhatten, NY: Rodale.

Davies, A. (2013). Identity and the assemblages of protest: The spatial politics of the Royal Indian Navy Mutiny, 1946. *Geoforum*, 48, 24–32.

Demaria, F. (2016). Can the poor resist capital? Conflicts over 'Accumulation by Contamination'at the ship breaking yard of Alang (India). In N. Ghosh, P. Mukhopadhyay, A. Shah and M. Panda (Eds.). *Nature, Economy and Society*. New Delhi: Springer, pp. 273–304.

Dolven, B., Elsea, J. K., Lawrence, S. V., O'Rourke, R. and Rinehart, I. E. (2015). Chinese land reclamation in the South China Sea: Implications and policy options. *Current Politics and Economics of Northern and Western Asia*, 24(2/3), 319–351.

Gilroy, P. (1993). *The Black Atlantic: Modernity and Double Consciousness*. London: Verso.

Hau'ofa, E. (1998). The ocean in us. *Contemporary Pacific*, 10(2), 392–410.

Helvarg, D. (2010). *50 Ways to Save our Oceans*. Novato, CA: New World Library.

Holbech, H. and Pedersen, K. L. (2018). Ballast water and invasive species in the Arctic. In N. Vestergaard Brooks, A. Kaiser, L. Fernandez and J. Nymand Larsen (Eds.). *Arctic Marine Resource Governance and Development*. New Delhi: Springer, pp. 115–137.

Honeyborne, J. and Brownlow, M. (2017). *Blue Planet II*. London: BBC Books.

Jay, S. (2008). *At the Margins of Planning: Offshore Wind Farms in the United Kingdom*. Farnham: Ashgate.

Jones, R. (2016). *Violent Borders: Refugees and the Right to Move*. London: Verso Books.

Lamers, M., Eijgelaar, E. and Amelung, B. (2015). The environmental challenges of cruise tourism. In C. Michael Hall, S. Gossling and D. Scott (Eds.). *The Routledge Handbook of Tourism and Sustainability*. Abingdon: Routledge, pp. 430–439.

Langewiesche, W. (2001). *The Outlaw Sea*. London: Granta Books.

Lebreton, L., Slat, B., Ferrari, F., Sainte-Rose, B., Aitken, J., Marthouse, R., Hajbane, S., Cunsolo, S., Schwarz, A., Levivier, A., Noble, K., Debeljak, P., Maral, H., Schoeneich-Argent, R., Brambini, R. and Reisser, J. (2018). Evidence that the Great Pacific Garbage Patch is rapidly accumulating plastic. *Scientific Reports*, 8, 4666 [[Online]]. doi:10.1038/s41598-018-22939-w (Accessed 22March2018).

Lehman, J. (2013). Relating to the sea: Enlivening the ocean as an actor in eastern Sri Lanka. *Environment and Planning D: Society and Space*, 31(3), 485–501.

Mack, J. (2013). *The Sea: A Cultural History*. London: Reaktion.

McLean, R. and Kench, P. (2015). Destruction or persistence of coral atoll islands in the face of 20th and 21st century sea-level rise? *Wiley Interdisciplinary Reviews: Climate Change*, 6(5), 445–463.

Peters, K. (2012). Manipulating material hydro-worlds: Rethinking human and more-than-human relationality through offshore radio piracy. *Environment and Planning A*, 44(5), 1241–1254.

Peters, K. (2015). Drifting: Towards mobilities at sea. *Transactions of the Institute of British Geographers*, 40(2), 262–272.

Rediker, M. (2007). *The Slave Ship: A Human History*. London: John Murray.

Ryan, A. (2012). *Where Land Meets Sea: Coastal Explorations of Landscape, Representation and Spatial Experience*. Farnham: Ashgate.

Singer, C. (2014). *Sea Change: The Shore from Shakespeare to Banville*. Amsterdam: Rodophi.

Steinberg, P. (2001). *The Social Construction of the Ocean*. Cambridge: Cambridge University Press.

Steinberg, P. (2015). Foreword. In M. Brown and B. Humberstone (Eds.). *Seascapes: Shaped by the Sea*. Farnham: Ashgate, pp. xi–xiv.

UNEP. (2017). [Online] http://web.unep.org/about/(Accessed 22 March 2018).

Urry, J. (2014). *Offshoring*. London: Polity Press.

Van Dover, C. L. (2011). Tighten regulations on deep-sea mining. *Nature*, 470(7332), 31.

Vives, L. (2017). Unwanted sea migrants across the EU border: The Canary Islands. *Political Geography*, 61, 181–192.

Part One
Approaches and advances

2 Architecture and design

Between seascape and landscape: experiencing the liminal zone of the coast

Jacky Bowring, Nancy Vance and Mick Abbott

If you stare very hard at the coast you will find it does not exist. In Benoit Mandelbrot's seminal work 'How long is the coast of Britain?' (1967) the 'coast' is revealed as an infinite set of intimate imbrications, where finer and finer increments of measure produce greater and greater lengths of coast. As each section is zoomed in on, the intricacies of the coastline expand in terms of its linear dimension, as first bays, then coves and then clefts are mathematically described. Once 1:1 and even microscopic resolution is reached and the coastal edge negotiates boulders, individual pebbles and grains of sand, the quest for an accurate measure evaporates. The coastline is infinite, resulting in the paradox that there is in fact no coast at all – rather there is a permeable relationship between land and sea. As cultural geographer and architect Anna Ryan confirms, with reference to the work of Tim Robinson and his investigations into fractals, the 'concept of coast as line has thus been *demonstrated* not to exist. It can only exist as an abstraction' (2012, 31).

Not only is there a challenge to the binary of the edge in the horizontal perspective, there is also a need to embrace a vertical expansion. Anthropologist Tim Ingold (2011) points to how our understanding of ground has become approximated into a two-dimensional plane that spreads out across the earth. And while the land has been reduced to a two-dimensional plane, so too has the ocean, as in anthropologist Claude Lévi-Strauss' conception of the ocean as a place of '"oppressive monotony"' and '"flatness"' (in Peters and Steinberg, 2014, 124). In terms of the land, Ingold (2011) argues for an understanding of landscape's thickness, a depth that reaches into the soil and then up amongst the interstitial spaces of plants, and then out into the substance of the atmosphere upon which the flight of birds is supported. And as an expansive view of the ocean, Peters and Steinberg (2015) propose a 'wet ontology', a recognition of the dimensionality of the sea in both temporal and spatial realms, of its fluidity and its flux, its expanse and its depth. And as Elden puts it, 'Just as the world does not just exist as a surface, nor should our theorisations of it; security goes up and down; space is volumetric' (Elden, 2013, 49). There is thickness too in the structure of our bodies, holding the air in our lungs, standing vertically upon the earth's surface, diving into the sea. Edges as separation are an illusion. The boundaries between things, between land and sea, are indeterminate. Rachel

Carson poetically describes how this indeterminacy is not only spatial, but also temporal,

> [f]or no two successive days is the shore line precisely the same. Not only do the tides advance and retreat in their eternal rhythms, but the level of the sea itself is never at rest ... Today a little more land may belong to the sea, tomorrow a little less. Always the edge of the sea remains an elusive and indefinable boundary.
>
> (Carson, 1998 [1955], 1)

In this chapter we explore the fluid sea–land continuum from the perspective of landscape architecture, echoing the concerns voiced by Steinberg and Peters in their ambition to think of the ocean as not a passive place but one where all of the things which make it up are 'imagined, encountered, and produced' (Peters and Steinberg, 2015, 256). This resonates strongly with the challenge we are seeking for an expanded landscape architectural conception of the coast, unsettling the neat containers of the plan or map view, and the sectional view (which often characterise disciplinary work), to embrace and liberate the coastal zone. Indeed, the very name of the discipline – *land*scape architecture – underscores the challenges of shifting focus to the fluid zone of the coast. There has been a longstanding hegemony of landscape when it comes to considering environmental design. Further, landscape architecture, like other disciplines dealing with the earth's surface, has been complicit in imposing precise edges between water and land, despite the flux that characterises this zone. As landscape designers we, the authors, aim to push beyond the observational and classificatory into the prospective, enriching the imaginative scope of this fluid zone. We propose the term 'coasting', as a mobilisation of the coast, rather than a static noun, 'the coast'. Coasting is something which happens; it is mobile and fluid, responding to the changing conditions and balances. It is neither finite nor fixed as a condition, and instead there is a sense of active exchange and flux, becoming more vivid as sea levels rise. However, rather than accepting this dynamism as a battle for territory between the land and the sea, we propose coasting as invitational, cooperative, discursive. 'To coast' could therefore be defined as a verb, referring to the processes involved in the perception and production of that condition which simultaneously embraces the four-dimensionality of the ocean and of the land. 'To coast', then, offers a particular insight and approach to living and working with the sea.

Challenging binaries

The escape from binaries is an effort to escape the ontologies that once separated humans from their place. The subject–object split, the severing of the body and mind, was an Enlightenment project, most notably by the philosopher René Descartes, and drove a wedge between humans and everything else – the non-human. Challenging this binary, and reconnecting mind and body, subject and object, is an impetus for coasting. Rather than land-then-sea, coasting explores ideas of the melting together of the binaries, tipping them from their containers: locating landsea and sealand.

One of the key works in challenging the neat binaries of modernity was the essay 'Sculpture in the Expanded Field', by Rosalind Krauss (1979). Krauss considered how the categories of architecture, landscape and sculpture were becoming inadequate for the kinds of works being developed by artists like Mary Miss[1] and Robert Smithson.[2] These works were neither landscape, nor architecture, nor sculpture, and Krauss built up an array of possibilities through mapping them in a Klein diagram, a form of oppositional structure where binaries are mapped against each other in order to develop a further layer of conditions. In Krauss' use of the Klein diagram she mapped the binaries of landscape and architecture, and not-landscape and not-architecture, to develop four new forms: marked sites, site construction, axiomatic structures and sculpture. This array is an expression of the 'expanded field' – an expansion beyond the previously accepted categories, such as architecture and landscape, where everything was neatly in its place. Krauss' work underpins the kinds of questions we are asking here, as we seek to unsettle the containers of sea and land, and to imagine different categories, innovative expressions and fluid conceptions.

The dissolution of binaries has been underway in cultural studies for the past few decades, including Edward Said's challenging of containers of 'The East' and 'The West' (1978), Donna Haraway's disruption of body and mind (1991) and William Cronon's unpicking of the culture/nature divide (1995), and for the discipline of landscape architecture, Krauss' work was extended by theorist Elizabeth Meyer in her challenging of the binaries of culture and nature, city and country, public and private, man and nature, landscape and architecture, what she refers to as 'false dichotomies and hierarchies' (1997, 45). Meyer asks the question, 'Has the binary thinking blinded us from seeing complex webs of interrelationships?' (1997, 45). Rather than land and sea as binaries, where each is defined by what it is not, we embrace the coast as a threshold that is at once of the sea and of the land. Rather than coast as edge, we may think of coast as a field that operates in between land and sea. As Anna Ryan states, '[d]efining an exact point when land becomes sea, or when sea gives way to land, is very difficult' (2012, 9). Rather than a binary condition of either/or, sea or land, the coastal zone is a both/and condition.

In discussing the aspiration for a more fluid understanding of the relationship between land and sea Anna Ryan points to the definition by Kerry County Council, Ireland, of 'beach' which includes 'foreshore and every beach, bank, cliff, sands, sand dune and every area contiguous thereto together with the foreshore waters for a distance of 300 metres seaward from the low water mark' (Ryan, 2012, 250). This definition by the Kerry County Council underscores the transitional nature of beaches, and that they are not fixed entities, but rather neither sea nor land, sometimes solid but sometimes a little more liquid, and without a clear edge. However, despite this definition of the sea and land being each of the other, the actual activities of the council failed to reflect such a broad definition – where rock walls

> work to reinforce the coast as a line. They display a politics of preservation that focuses on an immediate economic infrastructure that is utterly abstract to the nature of the coast as physical entity. The thickness of the coast is forgotten in remedial, short-term measures that ignore the long-term flowing cycles of material flow between land and sea, flows that are essential to understand in order to successfully inhabit the space of encounter between land and sea.
>
> (Ryan, 2012, 251)

In practice then, how can we overcome an ambition to recognise a fluid and non-binary definition of the coast, whilst also exploring how to intervene in ways which do not contradict this?

Turning to Aotearoa New Zealand we find an evolution of the apprehension of the coast from thin line to thick line, beginning with a printed map of Abel Tasman's 1642 voyage. The map shows the distinct edge of Tasman's voyage where, on hitting land, he turns north with the land to the starboard side, describing the land from the sea as a slowly forming edge. In his circumnavigation over 100 years later, Captain James Cook created a picture of large-scale islands (Figure 2.1). While the land has some mountains inscribed, most is left blank – an emptiness that as a common trope of colonialism provokes further 'discovery' regardless of what environments and cultures already have their location there.

Historian Paul Carter (1999) writes of the tradition of coastal surveys, and how the coast represented a very distinctive and elusive phenomenon for mapping. As he notes, the coast 'as the site of demarcation brought otherness into the world' (Carter, 1999, 132). Carter describes how 'the coast was a pre-emptive clearing', that it 'was a space already extracted and differentiated from the uniformity of nature' (1999, 132). Of course such diagramming is imprecise, and in Cook's maps Stewart Island remains tentatively linked to what is now called the South Island, while the volcanic Banks Island finds itself separated (see Figure 2.1). Almost 100 years later, while in command of the *HMS Acheron* and appointed to prepare the first full hydrographical survey of New Zealand, Lieutenant John Lort Stokes resolved such ambiguities.

Sheet XII of the New Zealand series, being from 'Foveaux Strait to Rr. Awarua on the West Coast', comprehensively documents the complex coastal edge with its filigree of fiords and islands. Yet despite its accurate reading of the coast, and the depth soundings along it, little attempt is made at describing the terrain beyond a sense of the hilliness of the country directly flanking it. Rather than describing the land, Stokes' focus was to define its coastline, and his cartographic style works to bring attention to this feature. What results lacks Mandelbrot's fractal representation. The map fails to express a coast of ecological interrelationships between fauna, flora, seasons, climate and tides. There is no

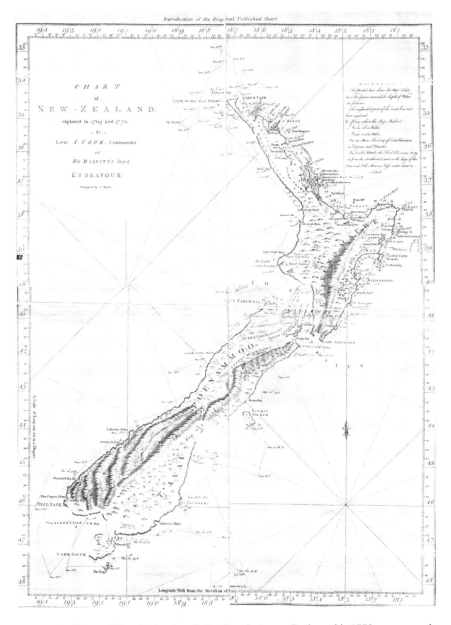

Figure 2.1 Chart of New Zealand made by Captain James Cook, on his 1770 voyage on the *Endeavour*. Image in the Public Domain, and is taken from the Project Gutenberg edition of *Captain Cook's Journal During the First Voyage Round the World*, which is based on an 1893 publication of Cook's original manuscript. Stewart Island is at the southernmost tip of 'T'Avai Poenamoo' (Te Wai Pounamu, what is now called the South Island), with a tentative link to the main island shown. On the east coast of 'T'Avai Poenamoo' an offshore island is shown, but this is in fact Banks Peninsula.

sense of the 'thick' interrelations that had preceded the settlers' activities in the south-west corner of New Zealand. For example, the archaeological record of Preservation Inlet reveals a coast rich with activities that join the ocean, coast, forest and communities as interconnected sites, and in which materials and food were harvested, and with which shelter and tools were also made. However, the nature of the coast created by Stokes is akin to a boundary rope set at mean high tide, where on one side is the sea made distinctive by various depth soundings while on the other side is a still uncertain terrain.

It is in this process of bounding of the land that Carter, in his spatial history of Australia, notes, 'the sea, formerly an asylum, itself becomes a prison, a turbulent, unavoidable barrier to progress' (1987, 34). Steinberg echoes this in pointing to how '[o]ceans have long been seen in Western thought as barriers' (2014, 14). Instead of a land hidden in a known ocean, the land becomes fixed and the ocean fluid. Or, as Carter phrases it, the 'sea yields to land; the sedentary replaces the dynamic' (1987, 35).

Yet in this reversal of positions, the coast still remains the bounded edge between land and sea. Carter notes also that understandings of the coast were founded on recognisable European-based typologies that included promontories, peaks, harbours, anchorages, estuaries, reefs and so on, these being 'uniform, dimensionless and self-repeating' (Carter, 1999, 127). In coastal maps, then, what made a harbour or reef significant was not that it was unique, but rather that it was like the other harbours and reefs on the chart. Ultimately, as Stokes completed his survey, the cartographically drawn coast of New Zealand peeled around on itself, until it was joined up as a series of three major islands (North, Middle and South) and a number of outliers.

For Stokes and other early cartographers, it was the land that was alien. This differs from contemporary apprehensions where a sense of being 'beyond' has often been directed at the sea. Anderson and Peters (2014, 5) point to how in Jackson's terms this contemporary position is one of 'negatively coding the ocean as "different" or "other"'. Alongside this spatial destabilisation is the recognition of the coast's temporality, as Waterton evokes of the wider non-representational framing, 'landscapes in this rendering are not static backdrops, but instead are imagined as fluid and animating processes in a constant state of becoming' (2013, 70). The recognition of places as *becoming* rather than existing in a definitive state is part of the challenging of the finite frame of binaries. Deleuze and Guattari see becoming as a 'deterritorialization' (1987), a term which is especially powerful in attempting to disrupt dualities dominated by land (*terra*). As noted by Dovey, relational thinking puts a different spin on concepts of place,

> [taking] the Heideggerean ontology of being-in-the-world [replacing it] with a more Deleuzian becoming-in-the-world. This implies a break with static, fixed, closed and dangerously essentialist notions of place, but preserves a provisional ontology of place-as-becoming: there is always, already and only becoming-in-the-world.
>
> (Dovey, 2010, 6, quoted in Anderson and Peters, 2014, 11)

Place-as-becoming offers a vivid conception of coasting, in its fluxing and morphing, conceptually and physically, and emphasises the realm of possibilities that such a re-framing presents.

Yet it is not only this binary that has proved problematic. Another of the binaries which has limited landscape architecture is that between architecture and landscape. This has been conceived of as the 'thick black line' (Solomon, 1988, 84) that is drawn around buildings, separating them from their context. We take inspiration from researchers and designers who have challenged this thick black line and revealed it not as a two-dimensional element, but as a place of richness in its own right (Assefa, 2003; Berrizbeitia and Pollak, 1999; Bishop and Bowring, 2001). Meyer's advice is that '[i]f we think of continuums or hybrids – of spaces in between – instead of opposing dualities, we do not have "others." If we do not have "others," we do not inherently value one term over another' (1997, 50).

For landscape architecture, the intervention of, for example, an engineered linear seawall is on the one hand a response which relies on a land–sea binary. Its role is to separate land and sea. Yet, on the other hand, there is a fluidity to the condition. While the wall appears immutable, the sea still moves up and down on it with the tide. Pushing the dissolution of binaries further encourages engagement with what Meyer calls the 'the relationships between things, not the things themselves' (1997, 66). Further, this exploratory mode engages with the ways in which materials and elements are arranged. How might the materials in this liminal space, this thick zone, be brought together, traversed, occupied, looked from or looked at? This is part of our aspiration to unsettle the containers which have characterised the coast, and to look at the tools in the landscape architect's toolbox in new and different ways.

Interventions and imaginings in coasting

As noted, typically landscape architecture has been complicit in the cleaving of sea from land, as the engineering underpinnings of the profession seek rational and often mechanistic solutions for how to join the land to the sea, or how to protect the land from the sea. But just as it has been recognised that the boundary of architecture and landscape is an interstitial zone of richness, so too, we argue, is the boundary of land and sea. Landscape architecture's language of analysing and imagining the world is based on horizontal views (maps and plans) and vertical views (sections and elevations). This rational language, shared by other spatial disciplines such as geographers and planners, typically slices and dices the world into image-able elements, including zoning, cadastral boundaries, and the setting of national and state boundaries. However, while it does this, it also misses out a great deal – for example, all of the temporal, sensory, experiential and emotional qualities that are similarly available.

That said, the difficulty of representing such qualities is no reason not to enlist them. For those in the design profession, some kind of image-making is necessary as part of the move towards the realisation of such ideas. As Carolan (2008, 412 cited in Waterton, 2013, 68) explains, '[i]t is not that we cannot represent sensuous, corporeal, lived experience but that the moment we do so we immediately lose something. Representations tell only part of the story, yet they still have a story to tell, however incomplete'. This is reinforced by Anderson and Peters, who observe that 'representation can only take us so far in knowing water worlds' (2014, 8–9). And further, Steinberg observes, 'an ocean is more-than-representational. It is continually reconstructed through our encounters, but as we engage the sea our experiences are performed and internalized through articulations with pre-existing imaginaries' (2014, 23). The oceanic water-world of coasting therefore presents both a challenge and possibility for designers in their need to represent, to 'image' and to imagine. All of this must happen in ways which somehow maintain the sensuosity of coasting, rather than cleaving from it all of those things which representations so often deny.

There is a tension, therefore, in representations and the need to document and specify. In a sense, we are bound by drawings. Drawings are inevitably two-dimensional, static and tend to portray an element with a sense of permanence. This permanence reflects the landscape hegemony, a landscape-centric way of being. Yet water is almost 800 times denser than air; elements degrade faster in water, materials erode and joints fail as more energy is dissipated on structures in water than would be on land. This requires a philosophical shift in approaching how to intervene with the coast as landscape architects. We therefore suggest approaches based on coasting.

Often drawings are plan-based, offering a bird's-eye view of the intervention, sited in its context, usually joining existing site features or structures. Plans deny the understanding of structural components beneath the surface and thus the understanding of how an object of intervention relates to both land and sea. A cross-section or sectional-elevation representation, which slices through the object, offers a view inside – one which reveals insertion points and transitions between the mediums: air, land and water. This resonates with Anderson and Peters' assertion that,

> a move towards a fluid ontology of oceans is thus not to claim that water worlds are taken as a perfect and absolute *bounded* space to study, in opposition to the attention paid to the land. Rather, a fluid metaphysics alerts us also to the ways in which the land and air fluidly merge and mix with water worlds too.
>
> (Anderson and Peters, 2014, 11–12)

Like the thresholds between architecture and landscape, coasting can be thought of as where 'transformations begin, where exchanges between likely and unlikely things occur and where identities are declared' (Berrizbeitia and Pollak, 1999, 82). To explore the territory between land and sea, and to conceptualise, as Meyer describes, a 'spatial continuum that unites, not a solid line that divides' (1997, 74), we suggest four modes of intervention used on the coastline: delineations, projections, overlays and sutures. Rather than being instruments of classification, these are introduced as constructive elements, which physically and metaphorically afford an experience within this space, and allow us to begin to reveal the relationships between, and exchanges provided by, land and sea.

Delineations

A delineation marks an outline and divides one element or space from another, one side of the line identified as distinctly different from the other. This creates an explicit binary relationship between the two items. Delineations can be seen to reinforce linear divisions along the coast, and include interventions such as flood banks, seawalls, and even the regular arrangement of lifeguard towers along a beach. In the design realm of the landscape architect working with the coast, delineations are generally horizontal, and parallel with (and thus reinforce), the coastline. Many towns in Aotearoa New Zealand reflect the legacy of mid-century philosophies of spatial organisation, locating a rail line adjacent and parallel to the coast, and then a road inland and parallel to the rail line, and the town behind that. This leads to a multiplicity of infrastructural barriers between residents and the sea.

Yet delineating interventions need not be so explicit, static and definite. A series of spatial slices may also echo the line of the tide moving in and out. And just as the high tide line might become a zone of deposited debris, so too can the delineation be a blurred zone, bridging or breaking down the binary. For pedestrians using the waterfront zone there is a demarcation of materials, perhaps between the sand of the beach and an urban surface. Roberto Burle Marx's Copacabana Beach in Rio de Janeiro, Brazil, brings a coalescing of sea and land, a challenge to a binary split between the two. Waves come onto the land in the form of the mosaic paving pattern, at a scale beyond a human stride, a scale that speaks to an oceanic perspective (see Figure 2.2). There is a sense of what Bishop and Bowring call 'dissolving', a melting of the usual sequence of the linear arrangement of city, then footpath, then sand, then sea. There is instead a 'termination of a formal relationship' as formal streetscape dissolves into seascape (Bishop and Bowring, 2001, 35).

Projections

A second type of intervention is a projection of the land into the sea, or the sea into the land, and might include breakwaters, piers and floating jetties. A projection, where a secondary object protrudes or extends into a space, shapes a relationship of familiar into unfamiliar, of object and/in space. Projecting can be a means of making an inaccessible environment accessible, like a balcony protruding off a building, or an observation deck over a canyon. It is also a multi-scalar condition, as projections also happen with a peninsula extending into the sea, or an estuary extending into the land. And at the micro-scale a rounded pebble on the waterline projects into the sea yet is still partially placed on the beach. However, projections are never permanent or inert. They are constantly

Figure 2.2 Copacabana Beach, Rio de Janeiro, Brazil. Paving design by Roberto Burle
Marx. Photo by Mteixeira62, 2012. CC BY SA 3.0.

under the influence of the dynamic forces of wind and water. There is a
reciprocity of influence.

Projections challenge the sea–land binary by offering, we argue, a gradient of
experience. There is the end closest to land – safe, sheltered – and the end
surrounded by the sea – exposed, windy, cold, perhaps wet. The sense of
precariousness increases with the movement from land into sea and with it an
understanding that, for humans, being terrestrial creatures, there is then less
certainty as the ocean gets deeper, as it hosts more marine life and is further from
'safety'.

The condition of projection can be understood in both the plan view and the
sectional view. In plan view a jetty or pier is an extension or finger of the land,
generally rectilinear, accessed from land on one side and surrounded by water on the
other three. In cross-section, the surface of a jetty or pier is elevated above the
water's surface while the structural components and access steps or ladders break
through the water's surface. A projection is a division of above and below; it can
offer elevation above the surroundings, and can allow for movement across the surf
zone, several metres above the water, within the 'fully voluminous or spherical
qualities of space' (Peters and Steinberg, 2015, 251), vividly revealing that coasting
is not only a disruption of the binaries of land and sea, but also sea and air.

The form also allows vertical access to deeper water where vessels can
dock alongside, and ladders and stairs can be fastened. An example of such a
form is Aotearoa New Zealand's New Brighton Pier in Christchurch, a 300
metre-long structure, sitting 7 metres above high tide on twenty 1.5-metre
diameter concrete piles (Figure 2.3). The pier gives an elevated perspective

Figure 2.3 New Brighton Pier, Christchurch, New Zealand. Photo by Bernard Spragg, 2015. Image in the public domain.

across the Canterbury Plains to the Southern Alps and across Pegasus Bay to the Seaward Kaikoura range. Walking on the pier means moving out of a usual land-bound condition, into being suspended in the sky above the sea, confounding the boundary between the two. The pier is a place of becoming. It is an exemplar of coasting. As in Assefa's (2003, 13) classification of architectural moves, this is a kind of interpenetration where one realm moves into another.

A floating jetty offers an added experiential dimension, one where the physical properties of water begin to be experienced more acutely, rising and falling, constantly and rhythmically under influence of the tide. The binary line falls away, and there is a blurring; the floating jetty is participating in something profoundly temporary, the astronomically influenced phenomenon of the tides. The floating jetty acknowledges buoyancy and focuses attention on what Ryan (2012, 18) calls the 'geometrical properties of the coast ... the alternative experiences of surface and depth'. Where land meets sea the jetty reciprocates and challenges the 'two geometrical conditions of horizontality and verticality' (Ryan, 2012, 18).

While piers and jetties project solid into liquid and unsettle the edge, a dry dock projects the sea into the land. This incursion of the oceanic into the terrestrial exemplifies how the 'coastline' can become an infinitely imbricated edge, such that it is no longer an edge at all. The dry dock fills with sea water, so it becomes a kind of artificial oceanic limb, as part of the hybrid condition of the coast. A boat can therefore float within a seemingly terrestrial context, to enable it to be worked on. While dualities, binaries and polarities rely on thresholds for definition, when these thresholds are troubled in this way new conditions must be imagined. This resonates with the destabilising of the architectural and landscape boundaries, through the strategy of displacement, where an element from one realm displaces that from another realm, as a form of deterritorialisation (Bishop and Bowring, 2001).

At the same time as the horizontal relationships of sea and land are played out through projection, dramatic shifts in water level can also intervene. One vivid example of this is Robert Smithson's Spiral Jetty, in the Great Salt Lake in Utah (1970). Within two years of its construction Smithson's huge gravel spiral landform was submerged as a consequence of the changing water levels in the Great Salt Lake. However, it began resurfacing in 1994 (see Figure 2.4). Projection is therefore revealed as temporal, disappearing and reappearing, with the landform projecting up into the sky, or the water projecting over the land.

Another environmental art work to invoke the temporality of projection is Mary Miss' Greenwood Pond (1989–1997), an urban wetland art installation with a ramp that diverges from a path and enters the pond, where it transitions to a concrete-lined trough that allows visitors to descend until they are at eye level with the water's surface. Through the installation,

Figure 2.4 Spiral Jetty from Rozel Point, Utah. Environmental Art by Robert Smithson. Photo by Soren Howard, 2005. CC BY-SA 2.0.

shifting between overviews and cut-outs within the water surface, the individual visitor is able to trace an intimate view of the place while putting together a new understanding of how it operates visually and physically ... invoking and building upon layers of associations and memories which have collected over time.

(Miss, 2010, n.p.)

This echoes the sensation of the RO & AD Architects' Moses Bridge, in Halsteren in the Netherlands (see Figure 2.5). This 'negative' bridge is a projection of air into the water, a memory of a defence line from the seventeenth century, and a reminder of how the binary of water and air is also a fluid condition (Moses Bridge, RO&AD Architecten, 2011). Accordingly, while delineations concern interventions that mostly divide, projections begin to blur and intermingle with the realm of the coast.

Overlay

A third means of 'coasting' and disrupting the apparent division between categories is through overlaying one material with another, for example plaza surfaces,

Figure 2.5 Moses Bridge, Halsteren, the Netherlands. Designed by RO & AD. Photo by M Appelmans, 2015. CC BY SA 2.0.

land reclamation, and artificial reefs. Overlays involve a layered arrangement of materials, elements or movements, in this case predominantly horizontal layers, echoing the long level line of the water–sky horizon. Interventions which overlay mix between the fixed condition of permanent materials and the dynamic of coasting – of a place in motion – where the liquid world brings constant change. Overlaying brings a mixing of layers, effectively destabilising the land–sea binary. Layers can reveal and conceal. For example, a thin layer of water can magnify what is beneath it. With this comes changing conditions of exposing and protection. Overlaying can occur naturally, as at many beaches, which at high tides are a rocky shoreline and at low tide reveal sandy beaches – a temporal overlay.

As a designed intervention an overlay can involve artificial forms such as those at Oriental Bay in Wellington, New Zealand, a design by Isthmus which uses stacked square concrete blocks to form an artificial headland to protect the beach from erosion (see Figure 2.6). This form affords recreation, exploration and seating for humans and more-than-humans, such as seals, as well as offering support for seaweed and algae – a dining place for fish. It becomes an urban interface with the tide, with high tide revealing only a small land-like area, dry and accessible, while at low tide many more areas are revealed, in varying stages of becoming 'marine'. Intertidal creatures have the opportunity to emerge at low tide, while urbanites

Figure 2.6 Oriental Bay, Wellington. Designed by Isthmus. Photo by Tom Beard, 2008. CC BY-NC 2.0.

become immersed in the overlay of coast in the city, the smell of seaweed, the precariousness of a slippery algae-covered surface, the rise and fall of the tide.

Land reclamation is also a form of overlaying, creating new land by filling an area with boulders, cement or clay and soil to a desired level. This is again a multi-scalar form of intervention, with much of the Netherlands being created through reclamation. Dubai's Palm Jumeirah and World Island forms illustrate how reclamation creates land, undoing a neat land–sea binary. In Christchurch there has been a range of ways in which reclaiming has worked, including natural reclamation where a raising of the base of Christchurch's Avon-Heathcote Estuary as a result of the earthquakes of 2010 and 2011 meant the estuary is now shallower and has a changed ecological condition. As a designed physical intervention, Te Awaparahi Bay Reclamation is a Government-approved 10-hectare reclamation at Christchurch's Port of Lyttelton using clean earthquake demolition material.

Causeways are also overlays of the land into the sea. An example of such is the Holy Island of Lindisfarne, Scotland, which is only accessible to cars at low tide. Pilgrims cross along a pedestrian route marked with line of stakes. Mont Saint-Michel in France is also joined to the land by a causeway, creating a considerable challenge to the binary of land and sea as the tides vary dramatically (see Figure 2.7). The causeway sits above and within the vast sandy beach, tethering the island to the land.

Figure 2.7 Mont Saint-Michel, France. Photograph taken 2005. Image in the public domain.

Its conditionality is apparent in its constantly varying form (some of which may be at the scale of a sand grain), with its horizontal edges continuously fluctuating with the tides' ebb and flow, and its vertical state being a balance between the pressures of gravity (amplified through the pressing down of vehicles and pedestrians) and the wetness of the subsoil. The arabesques of wetness and dryness in the image provide a vivid impression of the causeway's contingency within this realm.

Beneath the water's surface, artificial reefs exemplify a further intervention through overlaying, reclaiming land using massive sandbag structures for the dispersion of wave energy, creating habitat for reef organisms or conditions for surfing. This overlaying of the seabed further extends the practice of coasting, the interweaving of the solid and the liquid. Artificial reefs are Haraway's cyborgs (1991), neither culture nor nature, something between the two, and something that is 'becoming' as a condition, where artificiality becomes an irrelevance to the organisms that colonise the reef.

Temporality is read through tidal change, or even through a change of state, as in the ice roads in northern Canada as described by Vannini and Taggart (2014) in an over-ice vehicular journey north along the Dempster Highway to Inuvik. The solid sea ice routes blur the boundaries between land and sea, solid and liquid, and provide a temporary condition of overlay (Vannini and Taggart, 2014). The deposit of sand by alongshore currents can also alter conditions, as at Caroline Bay in Timaru, New Zealand, where the sand shore 'line' is thickening as deposits shorten the bay and expand the beach. And beneath the sea, overlays such as sunken ships intervene, stretching the zone of coasting.

Suture

The final form of intervention is of suture, a stitching of land to sea, in bridges, slipways, ladders and artificial waterways. Sutures can afford us features of amphibious or avian creatures, allowing access into other realms. A suture might be as simple as worn tracks in sand dunes linking car park and surf, or a timber slipway from which a rowing dinghy or kayak is launched, or adhered limpets and chitons experiencing the lift and recess of each tide over their surface, crossing the air–water interface. Most simply we might think of a slipway, or boat ramp, made of reinforced concrete, or simply stone, as a kind of suture.

This form of intervention can be limited, or constrained, by the height of the tide if the ramp does not extend far enough for a very low tide. To launch, the boat is floated off the trailer; to retrieve, it is winched back on the trailer. The boat ramp, the boat-on-the-trailer, and then the boat-on-the-water, illustrate the overlap between interventions. The boat ramp is simultaneously a suture, a projection and an overlay, while the motile element of the boat shuttles between all of these conditions.

Navigation marks – such as lighted buoys – stitch vessels onto the land; the blinking lights form a bridge between a watery environment and a solid one, as a kind of navigational prosthetic, a visual suture. Abandoned jetty piles become a vertical suture, in and out of water, being revealed and hidden as the tide moves, like a sewing needle moving up and down. The piles initiate a spatial relationship, stitching earth to sky through water, embedded in the substrate of the earth. They are projections while being sutures, where the land projects into water and air through the proxy of the pile, and the air and water project down between the piles.

Sutures are found too in bridges, causeways, viaducts and tunnels, as well as pathways, steps and landings. An extreme example of a suture is the world's longest underwater tunnel, the Channel Tunnel, a 50km-long stitch that sews France and England together, beneath the sea, at a depth of 75m. At the other end of the spectrum, the exposed and rocky headland of Punta Pite in Chile has a series of ocean-side interventions including stone steps, landings and walls linking a path along, and through, and into the sea (see Figure 2.8). They create a series of spatial experiences, twisting between rocks where the path widens and narrows. There are no railings. The staircases appear and disappear, stitching together previously unreachable vistas and tidal pools. Designed in 2004–2006 by landscape architect Teresa Moller, the pathway reflects her inspiration as a channelling of the words from a famous Chilean poet, who described Chile as '"pure geography"' (cited in Martignoni, 2007, 78). The design 'surrenders to the power and beauty of the ocean' (Martignoni, 2007, 78). These built interventions seem to grow organically from the landscape, inviting land into sea, sea into land, stitching them together.

The interventions suggested here form the sketch of a primer on coasting. The categories of delineations, projections, overlays and sutures are identified in terms of their emphases, rather than implying they are mutually exclusive. In the voluminous and contingent realm of the coast, whether an intervention is an overlay or a suture may depend on where you are, and what perspective you are looking from. It is possible that an intervention is all of these things, and in so doing it contributes to the becomingness of coast.

Conclusions

Rachel Carson notes that,

> the edge of the sea is a strange and beautiful place. All through the long history of the earth it has been an area of unrest where waves have broken heavily against the land, where the tides have pressed forward over the continents, receded, and then returned.
>
> (Carson, 1998 [1955], 1)

Our proposed categories of intervention – delineation, projection, overlay and suture – seek to embrace this strangeness and beauty, destabilising the binary of coast, offering a sense of *coasting*. This is a welcoming of the idea of the coast as a place in motion, ever becoming; a strange and beautiful place that can be drawn into

Figure 2.8 Punta Pite, Chile, designed by Teresa Moller. Photograph by Andres Briones, 2005. CC BY-NC-ND 2.0

our discipline, rather than being marginal. Ingold (2011) argues that the environment and our living in it is an unfolding and never-ending conversation. In this, the coast is more than a series of cycles based on tides and seasons. Our ongoing interactions make it, and ourselves, into always incomplete emergent forms.

In addition, the shift to consider the more-than-human (Whatmore, 2006) also encourages a consideration of how we shift our anthropocentrism when 'coasting'. Rather than seeing more-than-human, sea-based others as being there for our observation – whale watching, seal colonies, penguin encounters – what happens if we cede this privileged position? We can put our design skills to work in aid of watery creatures, through the design of fish ladders, bat bridges, bird roosts and fish reefs. We can also look beyond 'animal' more-than-humans who are companions in coasting, to the sea itself.

For landscape architecture the coast is often approximated to the role of container. We have invited ourselves into watery worlds through the design of jetties and slipways, or specific attempts to keep the water at bay. Our collective actions, through climate change and sea-level rise, are reversing the mode of intervention. This is especially vivid in Christchurch, New Zealand, which experienced 50 years of sea-level rise 'overnight', as a consequence of the earthquakes of 2010 and 2011 (Copley et al., 2015). The sea is coming into the land, and with it the coast. Rather than retreating, the coast is thickening, and generating a greater range of relationships, a constantly expanding field. Some of these might be changes to concepts of land ownership, recognising that owning land in a dynamic environment is counterintuitive (Copley et al., 2015). Perhaps roads may become urban waterways, and the submerging landscape becomes a fresh way of thinking about housing, traffic, agriculture, recreation, all challenging the presumed immutability of land (Copley et al., 2015). Like Krauss' 'expanded fields' (1979), these disrupted and deterritorialised concepts offer opportunities for innovations, different ways of thinking about the landsea/sealand condition, creating new imaginings and prospects for experiencing these sites. We imagine the coast becoming less distinct, more indeterminate and increasingly interstitial. Through 'thinking volume' (Bridge, 2013, 57), 'thinking with the sea' (Peters and Steinberg, 2015, n.p.), we become immersed, submerged, engaged, with what is not simply an edge between two planar conditions but a dynamics of depth, breadth and mutability. Rather than a place of simple engineering, the coast holds the potential to be an expanded site of cultural and ecological creativity as opportunities for gathering food, moving habitat and diverse experiences are increasingly encountered by that never-ending conversation formed from the meeting and making of landsea and sealand: a place of coasting.

Notes

1 American sculptor Mary Miss exemplified the evolution of non-object sculptures, where the works were often spaces rather than things. Miss' work is the first invoked by Krauss in her article, referring to the 1978 piece *Perimeters/Pavilions/Decoys*, which includes a set of structures spread through a landscape, ranging from a subterranean space to a tower. The idea that this whole ensemble was a 'sculpture' demanded that the definition be expanded.

2 Robert Smithson is an American environmental artist, best known for his work Spiral
   Jetty (1969–70) in the Great Salt Lake in Utah, USA. This work, along with many of
   his others, challenged what sculpture is, drawing in aspects of landscape, and the
   dynamics and ephemerality of phenomena such as the ebbing and flowing of the salt
   lake over time.

References

Anderson, J., and Peters, K. (2014). *Water Worlds: Human Geographies of the Ocean.*
   Farnham: Ashgate.
Assefa, E. M. (2003). Inside and outside in Wright's Fallingwater and Aalto's Villa Mairea.
   Environmental & Architectural Phenomenology, 14(2), 11–15.
Berrizbeitia, A., and Pollak, L. (1999). *Inside Outside: Between Architecture and Land-
   scape.* Gloucester, Massachusetts: Rockport.
Bishop, S., and Bowring, J. (2001). Layering, displacement, dissolution: Mapping the
   spaces between architecture and landscape. *Critiques of Built Works of Landscape
   Architecture*, 6, 30–37.
Bridge, G. (2013). Territory, now in 3D! *Political Geography*, 34, 55–57.
Carson, R. (1998 [1955]). *The Edge of the Sea.* New York: Mariner.
Carter, P. (1987). *The Road to Botany Bay: An Essay in Spatial History.* London: Faber and
   Faber.
Carter, P. (1999). Dark with Excess of Bright. In D. Cosgrove (Ed.). *Mappings.* London:
   Reaktion Books, pp. 125–147.
Copley, N., Bowring, J., and Abbott, M. (2015). Thinking ahead: Design-directed research
   in a city which experienced fifty years of sea level change overnight. *Journal of Land-
   scape Architecture*, 10(2), 70–81.
Cronon, W. (1995). The trouble with wilderness; or, getting back to the wrong nature. In W.
   Cronon (Ed.). *Uncommon Ground.* New York: W. W. Norton & Company, pp. 69–90.
Deleuze, G., and Guattari, F. (1987). *A Thousand Plateaus: Capitalism and Schizophrenia*
   (Trans. B. Massumi). Minneapolis: University of Minnesota Press.
Elden, S. (2013). Secure the volume: Vertical geopolitics and the depth of power. *Political
   Geography*, 34, 35–51.
Haraway, D. (1991). *Simians, Cyborgs and Women: The Reinvention of Nature.* New York:
   Routledge.
Ingold, T. (2011). *Being Alive: Essays on Movement, Knowledge and Description.* Abing-
   don: Routledge.
Krauss, R. (1979). Sculpture in the expanded field. October, 8, 30–44.
Mandelbrot, B. (1967).How long is the coast of Britian? Statistical self-similarity and
   fractional dimension. *Science*, 156(3775), 636–638.
Martignoni, J. (2007). Cliff-hangers: Improbable stairs lead down to the surf on the Chilean
   coast. *Landscape Architecture*, 97(August), 78–82.
Meyer, E. K. (1997). The expanded field of landscape architecture. In G. Thompson and F.
   Steiner (Eds.). *Ecological Design and Planning.* New York: John Wile, pp. 45–79.
Miss, M. (2010). Greenwood Pond: Double site. [Online] http://marymiss.com/projects/
   greenwood-pond-double-site/(Accessed 13 September 2017).
Moses Bridge, RO&AD Architecten. (2011). ArchDaily. [Online] http://www.archdaily.
   com/184921/moses-bridge-road-architecten/(Accessed 13 September 2017).
Peters, K., and Steinberg, P. (2014). Volume and vision: Fluid frames of thinking ocean
   space. *Harvard Design Magazine*, 35, 124–129.

Peters, K., and Steinberg, P. E. (2015). A wet world: Rethinking place, territory, and time. *Society & Space* Open Site [Online] http://societyandspace.org/2015/04/27/a-wet-world-rethinking-place-territory-and-time-kimberley-peters-and-philip-steinberg/(Accessed 15 September 2017).

Ryan, A. (2012). *Where Land Meets Sea: Coastal Explorations of Landscape, Representation and Experience*. Farnham: Ashgate.

Said, E. (1978). *Orientalism*. New York: Pantheon.

Solomon, B. S. (1988). The thick black line. In B. S. Solomon (Ed.). *Green Architecture and the Agrarian Garden*. New York: Rizzoli, pp. 84–85.

Steinberg, P. E. (2014). Mediterranean metaphors: Travel, translation and oceanic imaginaries in the 'New Mediterraneans' of the Arctic Ocean, the Gulf of Mexico and the Caribbean. In J. Anderson and K. Peters (Eds.). *Water Worlds: Human Geographies of the Ocean*. Farnham: Ashgate, pp. 23–37.

Vannini, P., and Taggart, J. (2014). The day we drove on the ocean (and lived to tell the tale about it): Of deltas, ice roads, waterscapes and other meshworks. In J. Anderson and K. Peters (Eds.). *Water Worlds: Human Geographies of the Ocean*. Farnham: Ashgate, pp. 89–102.

Waterton, E. (2013). Landscape and non-representational theories. In P. Howard, I. Thompson and E. Waterton ((Eds).). *The Routledge Companion to Landscape Studies*. London: Routledge, pp. 66–75.

Whatmore, S. (2006). Materialist returns: Practising cultural geography in and for a more-than-human world. *Cultural Geographies*, 13(4), 600–609.

3 Marine spatial planning

Sea Change Tai Timu Tai Pari: reflections on marine spatial planning in the Hauraki Gulf

Raewyn Peart

The Hauraki Gulf, located on the north-east coast of the North Island of New Zealand, has attracted human settlement for close to 800 years. Eastern Polynesians were the first settlers, and over subsequent generations Māori cultural practices developed in close association with the sea. The signing of the Treaty of Waitangi by representatives of northern Māori tribes and the British Crown in 1840, was the basis on which Britain asserted sovereignty over the North Island of New Zealand, with the more sparsely settled South Island claimed by virtue of discovery. The Treaty guaranteed (amongst other things) that Māori would have full, exclusive and undisturbed possession of their lands, estates, forests, fisheries and taonga (precious things).

Organised European settlement of the Gulf commenced with the establishment of the first capital of the new colony there in 1840. This started a process of fundamental change to the biophysical and ecological characteristics of the Gulf, with ongoing decline in the ecological health of the marine area that is continuing today. European settlement also started a process of Māori dispossession of their land, marine estate, fisheries and other cultural treasures resulting from serious breaches of the terms of the Treaty of Waitangi which has led to redress settlements in recent years.

Sea Change Tai Timu Tai Pari was an innovative three-year project, aimed at developing a marine spatial plan to help support healthy marine ecosystems as well as provide for cultural, social and economic engagement. The Māori phrase 'Tai Timu Tai Pari' in the project's title refers to the ebb and flow of the tide. The plan was designed to be bold and innovative, recognising that 'business as usual' had failed to stem ongoing maritime degradation. It was the first, and so far only, marine spatial plan to be developed in New Zealand.

Reflecting Crown obligations under the Treaty of Waitangi, and potential future Treaty settlement arrangements for the Hauraki Gulf, the project was overseen by a co-governance entity which had equal representation from Mana Whenua (Māori groupings with ancestral ties to the Hauraki Gulf) and government agencies. The plan itself was developed by a group comprising representatives from Mana Whenua and a range of sectoral groups. The members engaged in a collaborative process which successfully generated consensus on the contents of the final plan. Most of the members had deep personal and/or cultural connections with the Gulf, which likely contributed to the successful outcome of the project.

This chapter outlines the historical context for the marine spatial planning (MSP) process, the plan development process and the outcomes. It further explores some of the likely challenges arising for implementation of the plan. Finally, based on direct involvement in the Hauraki Gulf and the plan-making process, the chapter teases out some of the personal dynamics of the project. The focus of the Sea Change Tai Timu Tai Pari project on the Hauraki Gulf is particularly pertinent to the themes in this book, and permits a further understanding of how we may better 'live with' the sea. It also helps us to reflect more broadly on the utility of MSP processes as a mechanism through which a deeper engagement with the sea can be facilitated in a variety of places, helping to generate better understanding and more effective management.

Historical context of the Hauraki Gulf

The Hauraki Gulf is a shallow coastal sea comprising some 13,900 km^2 of waterspace which was created through the flooding of river valleys after the end of the last ice age (see Figure 3.1). It encompasses numerous islands, harbours and embayments. The Gulf is a highly productive marine system with some of the richest phytoplankton levels in New Zealand (Zeldis et al., 2004). The inner Hauraki Gulf is a major spawning and nursery area for snapper (*Chrysophrys auratus*) and other finfish (Zeldis and Francis, 1998) and it supports some of the most important inshore commercial and recreational fisheries in the country.

The Hauraki Gulf is also of significance in respect of its biodiversity. It is a 'globally significant seabird biodiversity hotspot', with over 70 species being sighted in the area, and 23 breeding there (Gaskin and Rayner, 2013, 5). The Firth of Thames contains a Ramsar wetland site which is part of the international flyway for migratory wading birds that arrive from northern Alaska and Russia each year. The area also supports nationally threatened bottlenose dolphins and Bryde's whales, as well as numerous common dolphins (Dwyer et al., 2014; Constantine et al., 2015). Te Hauturu-o-Toi (Little Barrier Island), situated in the north-west of the Gulf, is recognised as one of the most important nature sanctuaries in the country and possibly in the world. The island is co-governed by Ngāti Manuhiri and the Crown and has more coastal bird species than any other island in New Zealand (Department of Conservation, 2017).

The Hauraki Gulf may have been one of the earliest places in New Zealand to be settled by eastern Polynesians, possibly around 1300 AD. The earliest dated sites in the Gulf, of around this time, are located on the eastern side of the Coromandel Peninsula. However, the archaeological history of the Hauraki Gulf has, as yet, been poorly investigated so little is known about early settlement patterns (Furey, 1997).

New Zealand sea lions and fur seals were prolific in the Gulf prior to settlement, but due to early harvesting for food, they were eliminated from the region by around 1500 AD (Smith, 1989). Finfish and shellfish subsequently formed a major part of the Hauraki Gulf Māori diet. Harvesting was carefully managed to ensure the ongoing health and productivity of the stock. This

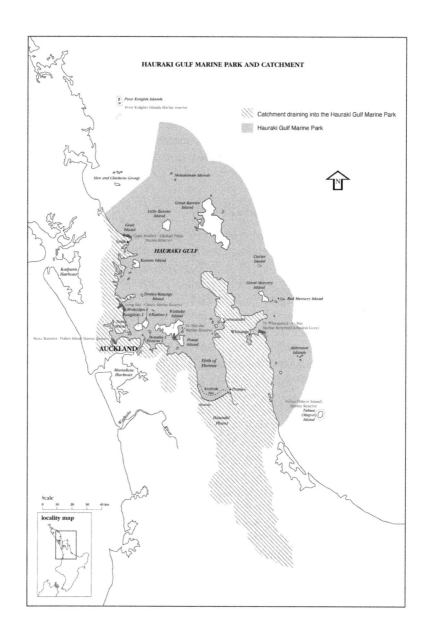

Figure 3.1 Map of Hauraki Gulf.

management was informed by Mātauranga Māori, described as 'the unique Māori way of viewing themselves and the world, which encompasses (among other things) Māori traditional knowledge and culture' (Waitangi Tribunal, 2011, 6). In the Māori world view everything is connected and interdependent, with mauri being the life force that gives things being and form. Marine management is focused on 'protecting and enhancing the mauri ora (life force) of fishing grounds and harbours' (Salmond, 2017, 276).

The first Europeans to visit the Hauraki Gulf were aboard Captain James Cook's vessel the *Endeavour*, which visited Mercury Bay and then the Firth of Thames in 1769. Cook was impressed with the abundant tall and straight kahikatea trees along the banks of the Waihou River which appeared suitable for spars (Reed and Reed, 1951). Reports of the trees on his return to England attracted the interest of spar ships which started visiting the area from 1794. However, there was little European interest in settling the Hauraki Gulf until 1840, when Englishman Captain William Hobson identified the Tāmaki isthmus on the Waitematā harbour as the location for the first capital of New Zealand.

Māori had long referred to the area as Tāmaki Whenua (contested land) or Tāmaki Makaurau (Tāmaki of a thousand lovers), referring to its history of being vigorously fought over due to its abundant natural resources and strategic location at the junction of two coasts and major river routes (Reed, 2001, 76). Ngāti Whātua had invited Hobson to locate the new capital on the isthmus in an attempt to create an enduring peace for the area. Cook subsequently renamed the area Auckland after his English patron Lord Auckland (Stone, 2001, 185–186; Peart, 2016, 68–73).

Auckland has now developed into the predominant urban centre in New Zealand, and by 2016 it had 1.6 million residents, a third of the national population (Statistics New Zealand, 2017). This large population concentration has put considerable pressure on the Hauraki Gulf in terms of the development of physical harbour works and reclamations, discharge of solid and liquid wastes and harvest of marine life. Indeed, early European interest in the catchments of the Hauraki Gulf – was largely focused on resource exploitation – kauri timber, but also flax and minerals such as copper and gold were extracted. After extraction, most remaining indigenous vegetation was crushed, burnt and replaced with exotic grasses suitable for grazing. The removal of indigenous vegetation cover resulted in the accelerated erosion of soils, thereby increasing the volume of sediment entering the Hauraki Gulf by around five times (Green and Zeldis, 2015). Increased sediment loadings have likely caused the loss of shellfish and seagrass beds, a reduction in fish productivity, and a general loss of marine biodiversity in the inner Gulf (Morrison et al., 2009).

However, one of the most significant ecological transformations was the drainage of the extensive wetlands in the Hauraki Plains, with around 50,000 hectares of wetland converted to farmland (Waitangi Tribunal, 2006). This was accompanied by extensive river works to reduce flooding, with the major river draining into the Gulf, the Waihou, being shortened by seven kilometres (Watton, 1995). The works allowed water to travel down the river at speed during floods, thereby reducing the risk of overtopping banks, but channelled large amounts of sediment, as well as nutrients from dairy farming, directly into the Firth of Thames.

The other major pressure on the natural environment of the Gulf has been fish harvesting. The Gulf was at the heart of the initial development of the commercial fishing industry in New Zealand during the late 1800s, initially driven by the needs of the growing urban population. By the late 1880s, around 40 commercial fishermen were engaged in fishing the Gulf with hand lines, baited hooks and beach seine nets (Paul, 1977). These small-scale methods had little immediate impact on fish stocks, which appeared to the European settlers to be unlimited.

Steam trawling was first introduced to the Hauraki Gulf in 1899, the first such trawling to be undertaken in New Zealand. It had an immediate impact on the catches of the other fishermen, who were successful in persuading the government to introduce a trawling ban for the inner Gulf in 1902. As the population grew and demand for fish increased, the ban was subsequently lifted, and trawling resumed in 1915. Other bulk methods of fishing were gradually introduced, including long-lining in 1912 and Danish Seining in 1923 (Paul, 1977). All these methods are still used today. One of the long-term impacts of using fishing methods that drag gear across the seabed has been the wide-scale loss of three-dimensional biogenic habitats in the Gulf, such as horse mussel beds, bryozoan outcrops and sponge gardens. These, scientists later discovered, were very important for the survival of juvenile fish, amongst other things (Morrison et al., 2014). Moreover, the Firth of Thames and much of the inner Hauraki Gulf was once covered with thick green-lipped mussel beds. During the late 1920s around three boats were dredging for mussels full-time, with peak landings made in 1961 when 2,800 tons of mussels were harvested. The stocks quickly collapsed and the fishery closed in 1969 (Paul, 2012). The rich mussel beds had been largely destroyed and they have never recovered.

Integrated management of the Hauraki Gulf

Hauraki Gulf Maritime Park

The concept of managing the Hauraki Gulf as a single entity had its conception in the establishment of the Hauraki Gulf Maritime Park, with the associated Hauraki Gulf Maritime Park Board, in 1967 through special legislation. Some-what surprisingly, although the Park was described as "maritime", it solely consisted of coastal land and excluded the sea. This land focus reflected the genesis of the Park initiative, which was concern that islands in the Gulf were being alienated from public ownership and public access to them was being restricted (O'Brien, 1971). The Park started modestly with just 4,926 hectares of land, but by early 1971 an additional 3,588 hectares had been added through a combination of transfer, purchase and gift. Much of the new land, such as that used for lighthouse operations, was already owned by the Crown but had been previously managed by other public agencies. Māori tribal groups generously gifted a number of islands including the whole Alderman group. Once all suitable land in public ownership was passed over to the Board, the rate of land acquisition slowed abruptly (Hauraki Gulf Maritime Park Board, 1968). But accumulating 27 reserves under the mantle of a maritime park, including many islands, was a major achievement and an enduring legacy for future generations.

During the 1970s and 1980s, there was a growing awareness in New Zealand of the need to extend conservation measures from the land into the sea, illustrated by the creation of New Zealand's first marine reserve near Leigh in 1975 followed by a marine park at Tāwharanui in 1981 (Peart, 2016). The Hauraki Gulf Maritime Park Board was disbanded in 1990 as part of a broader government restructuring exercise, and its function of managing the islands was taken over by the newly established Department of Conservation. During the 1980s, a former Board member, Allan Brewster, lobbied for the establishment of a marine park in the inner gulf, where commercial fishing would be excluded. This failed to get political traction at that time, but in 1990 the National Party's spokesperson on conservation announced an intention to promote a complete marine park for the Gulf (Waitangi Tribunal, 2001). This was followed by various other initiatives, but it was not until 1998 that the Hauraki Gulf Marine Park Bill was introduced to Parliament. It was finally passed into law in February 2000. Brewster's recreational fishing park proposal for the inner Gulf later re-emerged in 2014 as one of the pre-election promises of the National Party, but again failed to gain traction.

Hauraki Gulf Marine Park

The Hauraki Gulf Marine Park, established under the new legislation in 2000, took a broader approach than the earlier maritime park concept. Instead of focusing largely on the islands (and solely on land), the new Marine Park included the seabed and seawater as well as reserves and conservation land. The boundaries of the new Marine Park were more constrained than the maritime one, following the marine jurisdiction of the Auckland and Waikato regional councils and excluding islands further to the north (which were under the jurisdiction of a separate regional council), including the Poor Knights and Hen and Chickens Islands.

During the development of the legislation, there had been some debate about the appropriate boundaries of the new Marine Park. Differing views were expressed, including by those who argued for a small park encompassing only the inner islands, which could be seen and easily accessed from the mainland (a 'land-based' view), and those who argued for an expansive park taking in all the islands and marine waters within the councils' jurisdiction (a 'sea-based' view). Regional council politician Mike Lee, who had worked as a ship's officer transiting to and from Auckland, observed that my time at sea was probably one of the reasons that I argued that Cuvier Island should be included in a more extensive marine park, as it was a mariner's signpost into and out of Auckland when sailing to and from the Pacific. After passing the Cuvier light, we would watch the land slip away over the horizon and only then we felt we were out of the Hauraki Gulf and at sea.

(quoted in Peart, 2016, 88)

This broader sea-based view ultimately prevailed.

The legislation laid out purposes for the Marine Park, including recognising and protecting its international and national significance; recognising the special relationship of tangata whenua with the Park, and sustaining its life-supporting capacity. Unlike other marine parks, however, no restrictions were explicitly placed on any activity, although a common set of management objectives were to apply to statutory decision-making within the Park (Hauraki Gulf Marine Park Act 2000, section 33).

The legislation also established the Hauraki Gulf Forum, a new entity to oversee the management of the Hauraki Gulf and its catchments. Instead of having direct management responsibilities, the Forum was conceived as an integrating body, bringing together the myriad of agencies that now played a role in the Gulf, so its members largely consisted of central and local government representatives. In addition, six tangata whenua representatives were appointed by the Minister of Conservation, thereby recognising the strong cultural linkages between numerous Māori tribal groupings and the Gulf.

An independent review of the performance of the Forum in 2015 identified some notable successes, including the preparation of guidance material on the interpretation and application of the Act's objectives and production of State of the Gulf reports. The Forum has also led a range of communication initiatives and public forums 'thereby socialising the "call to action" … and the nature of the environmental challenge' (Bradly, 2015, 14). In addition, the Forum was very influential in the formation of the Sea Change Tai Timu Tai Pari project described below.

The review also found significant shortcomings in the effectiveness of the organisation. This included members being unable or unwilling to consistently work together as a 'political peer group' to consistently promote the objectives of the Act, a lack of clarity of purpose, a perceived politicisation of science and state of the environment reporting, a lack of 'teeth' and insufficient resourcing (Bradly, 2015, 15–16). The Forum employs only one staff member and operates on a small budget of around NZ$255,000 (Hauraki Gulf Forum, 2015, 9).

The report compiled a number of recommendations on how the current situation could be addressed, including that '[g]overnance should be reformed and the current structure replaced with a smaller, more agile Forum membership that provides a peer group of politically aware and strong leaders committed to promoting the objectives of the Act' and that it 'needs greater representation of tangata whenua to reflect the nature of the Crown-Iwi partnership' (Bradly, 2015, 5).

State of the Gulf report

One of the significant functions of the Forum was to prepare a state of the environment report on the Hauraki Gulf every three years. It was the Forum's 2011 report which effectively brought to public notice the dire ecological state of the Gulf and laid the foundations for the Sea Change Tai Timu Tai Pari project. The report concluded:

> This report highlights the incredible transformation the Gulf has undergone over two human lifespans. That transformation is continuing in the sea and around the coast, with most environmental indicators either showing negative trends or remaining at levels which are indicative of poor environmental condition. It is inevitable that further loss of the Gulf's natural assets will occur unless bold, sustained and innovative steps are taken to better manage the utilisation of its resources and halt progressive environmental degradation.
>
> (Hauraki Gulf Forum, 2011a, 13)

One of the ground-breaking features of the report was that it compared the current situation with what the Hauraki Gulf would have been like pre-European settlement. This was to overcome the 'sliding baseline' phenomenon[1] that occurs when people compare what they see today with what they remember during their lifetimes and where, as generations move on, the current degraded situation becomes the new norm or 'baseline' (Gatti et al., 2015). It is clear that there has been profound change over the past generations. Early European explorers recorded abundant marine life in the Gulf, including Captain Dumont D'Urville who wrote about his experiences being becalmed in the Tamaki Strait in 1827:

> Whenever we are becalmed, the crew immediately catch with their lines an amazing quantity of splendid fish, belonging to the dorado group [snapper], which are delicious food ... it seems to be extraordinarily abundant in these water. While we were anchored in the mouth of the river Magoia [Tāmaki River], the Tāmaki natives filled their canoes in a few hours. Today the crew soon caught hundreds of them and they had enough for each mess to be able to salt down a good stock.
>
> (Wright, 1950, 167)

Even during current lifetimes there have been marked changes. For example, Darren Shields, who began diving at the age of five during the 1970s and is one of New Zealand's top spearfisherman, recently observed:

> when I first started diving, in any given spot you would see 50 big kingfish, weighting 20 to 25 kilos each ... You still see them but they are smaller. You never see bulk large fish any more. There were places where you would see hundreds of John dory and tarakihi in a day and now you see none.
>
> (quoted in Peart, 2016, 2011)

Seachange Tai Timu Tai Pari

Inception of project

As already indicated, the Sea Change Tai Timu Tai Pari project had its inception in the Hauraki Gulf Forum's 2011 State of our Gulf Report which indicated that current management approaches were not sufficient to reverse the ongoing environmental decline of the marine system. At the same time, there was growing awareness that an approach called 'marine spatial planning' (MSP) was being increasingly applied in other jurisdictions. For example, Jay, Ellis and Kidd (2012) describe the early take-up of MSP in Australia followed by formal MSP institutional arrangements being put in place in North America, northern Europe and the western Pacific. Notably, the United Kingdom made explicit legal provision for MSP in the Marine and Coastal Access Act 2009.

MSP is a rational and strategic approach which can be used to proactively plan for the future use of the marine environment. At its heart is a concern to protect the underlying ecological backbone or productivity of the marine area, but it also seeks to reduce conflict and maximise synergies, providing greater certainty on where marine activities can and cannot locate (Interagency Ocean Policy Task Force, 2009; Ehler and Douvere, 2009).

In order to understand what such an approach might contribute to the Hauraki Gulf, I was commissioned by the Forum to undertake an international review of MSP. This was intended to distil key approaches and learnings, and to identify how the approach might be applied to the Hauraki Gulf. The review investigated planning processes in Australia (Great Barrier Reef and Common-wealth waters), the USA (Florida Keys Marine Sanctuary, Massachusetts state waters and Rhode Island state waters), Canada (Eastern Scotian Shelf) and Norway (Barents Sea-Lofoten Islands). The report, which was released in 2011, concluded that '[m]arine spatial planning is a well-accepted strategic planning process which could help achieve the purposes of the HGMPA [Hauraki Gulf Marine Park Act] including integrated management and the protection and enhancement of the life-supporting capacity of the Gulf' (Hauraki Gulf Forum, 2011b, 40).

The release of the report generated considerable interest, and with the encouragement of the Hauraki Gulf Forum and the Environmental Defence Society,[2] Auckland Council and the Waikato Regional Council agreed to lead a MSP project, with the Department of Conservation and Ministry for Primary Industries subsequently joining the agency grouping. Each of the agencies has a different role and interest in the Gulf and they jointly funded the planning process. The two regional councils are responsible for managing the catchments and most marine activities with the Marine Park (including aquaculture but excluding fishing activity). The Department of Conservation manages marine reserves and is the prime agency looking after protected species such as marine mammals and seabirds. The Ministry for Primary Industries directly manages fisheries and has a role in the promotion of agriculture, aquaculture and forestry.

The following sections of the chapter will reflect upon the process of formulating a marine spatial plan. The focus in MSP tends to be on the plan itself, without considering *how that plan came into being*. These reflections therefore enable a deeper understanding of decision-making processes that occur when we attempt to design how we might better live with the sea. The process for developing the plan was experimental, and it evolved as the project progressed through a 'learning-by-doing' approach. This is in line with the suggestion by Kidd and Ellis (2012) that trial-and-error experimentation, controlled risk-taking and long-term adaptation may be more appropriate in complex planning environments such as the sea. Novel elements of the process included a wide scope which embraced catchments as well as the sea, the use of a co-governance structure to oversee the process, and the adoption of a stakeholder-led collaborative process to develop the plan (see Figure 3.2).

Co-governance

As a result of Treaty of Waitangi settlements, Māori–Crown co-governance arrangements have been established for a range of natural resources in New Zealand, including the volcanic cones in Auckland, the Waikato River, Te Urewera (formerly the Urewera National Park) and the Whanganui River. In the case of the latter two, the resource was also declared a legal entity or person in its own right, a highly novel legal development in the New Zealand context (Beverley, 2015). The Gulf MSP process was complicated by the lack of resolution of Treaty claims for the Hauraki tribes who have a major interest in the Gulf. Part of their Redress Deed, only finalised in December 2016 after the

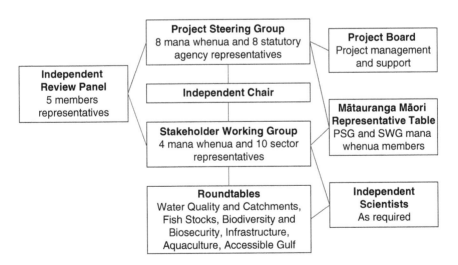

Figure 3.2 Organisational structure of the Sea Change Tai Taimu Tai Pari planning process.

planning process had been completed, acknowledges the aspirations of the Hauraki iwi for co-governance arrangements for the Gulf. In the Redress Deed, the Crown agrees to negotiate redress in relation to the Gulf 'as soon as practicable, and will seek sustainable and durable arrangements' involving the Hauraki iwi in the natural resource management of the Gulf based on the Treaty (Hako et al., 2017, 105–106). It is therefore possible that such future arrangements may involve the Hauraki Gulf becoming its own legal person with formalised co-governance arrangements replacing those of the Hauraki Gulf Forum.

Recognition of this aspiration of iwi, and ongoing obligations of the Crown under the Treaty, resulted in a co-governance arrangement being put in place to oversee the MSP process. This took the form of a 16-member Project Steering Group (PSG), which consisted of eight statutory body representatives and an equal number of Mana Whenua representatives. The PSG was led by two co-chairs, one from Auckland Council and a second from Mana Whenua. The PSG set the purpose of the project as being to,

> [d]evelop a spatial plan that will achieve sustainable management of the Hauraki Gulf, including a Hauraki Gulf which is vibrant with life and healthy mauri, is increasingly productive and supports thriving communities. It aims to provide increased certainty for the economic, cultural and social goals of our community and ensure the ecosystem functions that make those goals possible are sustained.
>
> (Sea Change Tai Timu Tai Pari, 2013, 1–2)

This highlighted the ecosystem focus of the plan, which was designed to address much more than the allocation of marine space to different users. The role of the PSG was to approve the terms of reference for the Stakeholder Working Group (SWG) and consider and adopt the final plan.

Collaborative processes

After some negotiation between the governance parties, agreement was reached to use a collaborative MSP process. This was to task a group of stakeholder representatives (called the Stakeholder Working Group (SWG)) with the plan development process, as opposed to the more usual process where an agency develops a plan in consultation with stakeholders. This approach accords with the suggestion by Kidd and Shaw that there is 'a particular need for innovative approaches to stakeholder engagement in marine planning contexts' (2013, 191). A collaborative process is one that uses consensus processes in order to reach agreement between the various parties. Holley and Gunningham describe it as 'a process where two or more stakeholders pool knowledge and/or tangible resources (e.g. information, money, labour) to reach agreement and solve a set of shared problems' (2011, 312).

The strength of the collaborative process is that it is designed to draw stakeholders closer together, so that agreement on the plan contents can be reached. Some of the potential benefits of collaboration include the generation of wiser and more durable decisions (through the use of shared information, a richer understanding of values and a problem-solving focus), the building of social capital (which reduces conflict and develops trust over time), and the promotion of change through participants sharing resources and changing their understandings and values (Yaffee, 2002).

On the other hand, collaborative processes have been criticised on the basis that they are undemocratic (because there is no direct democratic accountability); favour development over the environment (as development interests can wield greater power in the process); displace collaborative goals with process goals; and require participants to uphold decisions regardless of the outcome (Brower, 2016). That said, the use of collaborative processes to help resolve intractable and complex natural resource management issues has become more frequent in New Zealand over the past decade, with one of the most high-profile processes being the Land and Water Forum which was established in 2008 by the Environmental Defence Society to develop national freshwater policy. This collaborative body, involving 68 different organisations with an interest in freshwater, successfully reached agreement on some 156 recommendations to government (Brower, 2016). The Sea Change Tai Timu Tai Pari process was inspired by this earlier model.

The Hauraki Gulf marine spatial plan was developed by the SWG, which consisted of 4 Mana Whenua and 10 representatives from commercial fishing, recreational fishing, farming, aquaculture, infrastructure, community and environmental interests. I was an environmental representative on the group. The selection of the 10 stakeholder representatives started with a shortlist of 700 interested organisations and was undertaken through a series of public meetings and a facilitated voting process whereby each sector identified one representative, who was then scrutinised by other sectors, to ensure that the final SWG members were able to work collaboratively with a range of people (Independent Review Panel, 2012). The role of the SWG, as set out in the Terms of Reference, was to

> [c]ompile information and evidence, analyse, represent all points of view, debate and resolve conflicts and work together as a group to develop a future vision for a healthy and productive Hauraki Gulf. This includes identified preferences for the allocation of marine space. The future vision will be manifested as a physical document – the Hauraki Gulf Marine Spatial Plan.
> (Seachange Tai Timu Tai Pari, 2013, 2)

The group was to operate on a consensus basis, which means that 'every member either supports or does not actively oppose (can live with) the decision' (Seachange Tai Timu Tai Pari, 2013, 4). Apart from this broad purpose and approach, there was little guidance as to how the plan was to be developed or what process was to be followed. This provided a great deal of flexibility and enabled the process to be adjusted when problems arose, but it also created some uncertainty, particularly as to the role of the sponsoring statutory agencies as they were funding the process but were not formally represented on the SWG. The original timeframe for the project was 18 months, but this was later extended to three years once it became apparent that the timeframe was too short to enable a robust plan to be developed.

The SWG was assisted by an Independent Chair, appointed by the PSG, who chaired the SWG meetings and provided liaison between the SWG and PSG (Seachange Tai Timu Tai Pari, 2013, 4). The work of the SWG was overseen by an Independent Review Panel comprising five experts in various fields, including Paris-based Charles Ehler, who was the co-author of the UNESCO guide to MSP. The Panel provided three reports and many of the recommendations were adopted by the SWG.

The SWG first convened in December 2013 and it met approximately monthly over a three-year period, with a six-month break during mid-2015, when the considerable challenge of integrating Mātauranga Māori and the sponsoring agencies' viewpoints into a plan being developed by a group of stakeholders within a tight deadline threatened to derail the project. When tensions started to escalate, the PSG 'paused' the plan-making process, while consideration was given to how to get the project back on track. A new Independent Chair was brought on board, the project structure was simplified, more time was provided for the plan-making process and greater support was provided to the Mātauranga Māori component. This enabled the successful completion of the project and is an example of how using a flexible process, that can be adapted to address problems as they arise, can be important to the success of complex projects (as suggested by Kidd and Ellis, 2012).

During the early stages of the project six 'Roundtables' were established to focus the plan development work on key elements of the overall picture as well as to involve a broader range of stakeholders. These groups were co-chaired by SWG members and included others who represented a wider range of stakeholder interests. Members endeavoured to agree on a vision and problem definition and then focused on developing solutions to the problems identified. A six-month time frame was provided for their work. Several Roundtables scheduled field trips so that members were able to jointly get out into the field and interact in a more informal setting. This helped build up a common knowledge base and engender trust. However, the short timeframe for the operation of the Roundtables meant that, in most cases, only high-level solutions were developed and controversial issues, such as marine protection, were parked. That said, the reports prepared by the Roundtables formed the building blocks of the plan.

Use of science and indigenous knowledge

Early on in the process, the SWG members agreed that the plan would be based on existing scientific knowledge as there was not the time or budget to commission new work. Spatial data sets were assembled on a web-based tool called SeaSketch for easy accessibility. These included data on the marine environment (e.g. bathymetry, currents, substrate, marine habitats, species distribution, areas of high ecological value and ecosystems services), uses (e.g. commercial and recreational fishing, boating, shipping and aquaculture) and catchment condition (e.g. water quality, land cover and land use). The software enables users to test and communicate spatially referenced ideas. A technical team, consisting largely of staff seconded to the project from the participating statutory agencies, was assembled to support the SWG and topic Roundtables and to access science as requested. This was later refined to a core group focused on plan writing headed by an independent lead writer and two science advisors.

Several mechanisms were put in place to assist with incorporating Mātauranga Māori into the plan. A Mātauranga Māori Representative Table was established comprising Mana Whenua members of the PSG and SWG to focus on developing plan material in this area with specialist support. A Mana Whenua writer and spatial information expert were incorporated into the plan writing team to ensure effective integration of the material as the plan was developed. In addition, the plan structure and presentation was informed by a specialist Māori designer to encapsulate a Māori world view. However, reconciling indigenous and Western world views in a planning process did not prove easy and full integration was elusive. Despite this, several participants in the project considered that it was more successful in this regard than other planning processes undertaken in New Zealand. Mātauranga Māori was a significant and important component of the plan although it did not achieve equal footing with Western world views (Peart, 2017).

Public interaction

An extensive public process was undertaken alongside the SWG. This involved public meetings and events, 25 'Listening Posts', a web-based use and values survey and an active social media, website and email updating programme. Overall the process connected with more than 14,500 people. The results of the engagement were summarised and made available to the SWG members to inform plan development and quotes from public comments were included in the plan itself (Sea Change Tai Timi Tai Pari, 2017). In addition, the group of community stakeholders present at the meetings held to select the SWG members, called the 'Hauraki 100+', were convened every few months to provide an update on progress, discuss key issues and obtain feedback.

Plan contents

The final plan, which was agreed to by all SWG members, was handed over to the PSG on 9 November 2016, and was publicly launched a month later on 6 December. The plan is structured around four parts, or *kete* (baskets) of knowledge: Part One, Kaitiakitanga and Guardianship; Part Two, Mahinga Kai – Replenishing the Food Baskets; Part Three, Ki Uta Ki Tai – Ridge to Reef or Mountains to Sea; and Part Four, Kotahitanga – Prosperous Communities (Sea

Change Tai Timu Tai Pari, 2017). The front end of the plan consists largely of objectives and actions, and this is supported by a summary of the scientific basis underpinning the plan provided in the appendices. Some key elements of the plan are described below.

The fish stocks chapter (located in the Mahinga Kai section of the plan) was based on two broad strategies, first to apply an ecosystem-based approach to harvest management, and second to protect and enhance marine habitats, thereby increasing the ecological productivity of the Gulf. Restoration of marine habitats is to focus on a nested approach. Large benthic areas (and eventually the entire Marine Park) are to be protected through the retirement of seabed-impacting fishing gear. Smaller areas within these zones will be the focus of passive restoration (through the establishment of marine reserves) and active restoration through the transplanting of species or establishment of new habitat patches (Sea Change Tai Timu Tai Pari, 2017, 67–76). A novel proposal was the creation of 'Ahu Moana' co-management areas. These will be located in nearshore areas extending one kilometre seawards. They will be co-managed jointly by Mana Whenua and local communities, in order to help strengthen customary practices associated with the marine space, as well as more effectively control harvest levels, particularly in areas under increasing pressure from the growing Auckland population (Sea Change Tai Timu Tai Pari, 2017, 48–49).

The plan identifies 13 new aquaculture areas and 13 new protected areas as well as an extension in size of two existing marine reserves. Provision has been made for customary harvest to take place in these areas and for the adverse effects on commercial fishers to be addressed. In addition, there is to be a 25-year review of the protected areas, and co-governance and management of them once established. (Sea Change Tai Timu Tai Pari, 2017, 118–120). This represents a marked change in approach to the marine reserve model currently applied in New Zealand, where reserves are managed by the Department of Conservation, and are considered to be permanent fisheries exclusion areas. The novel approach in the plan resulted from the need, in a collaborative process, to accommodate all interests, including those of Mana Whenua and commercial fishers.

The impact of poor water quality on the ecological health of the Hauraki Gulf was one of the greatest areas of concern, with the main stressor being sediment. The plan identifies a number of actions to address this, including the development of catchment management plans (starting with four high-priority catchments), establishment of catchment sediment limits, increase of sediment traps through reinstating natural or engineered wetland systems, ensuring the adoption of good sediment practice by all land users and retiring inappropriate land use on highly erodible land (Sea Change Tai Timu Tai Pari, 2017, 130–137). Furthermore, nutrient enrichment was an emerging water quality issue in the Firth of Thames, with zones of oxygen depletion and acidification occurring seasonally. The plan places a cap on nitrogen discharge levels, which are to be kept at or below current rates, until sufficient scientific work had been completed to enable an appropriate nutrient load limit to be put in place (Sea Change Tai Timu Tai Pari, 2017, 141–142).

The plan also addresses the relationship of people with the Gulf and has as a key initiative 'inspiring the Hauraki Gulf Marine Park community', including connecting people with the place through the use of storytelling, supporting volunteer programmes, creating greater opportunities for people to access the Gulf (including walkways, coastal reserves and water-based transport) and supporting the expansion of marine education (Sea Change Tai Timu Tai Pari, 2017, 161–163).

Implementation

The initial response to the release of the plan was generally positive. The three key government ministers – Environment, Primary Industries and Conservation – 'welcomed' the plan, and indicated that Government would 'establish a process to formally consider the plan, before developing and consulting on recommendations on how to implement it' in a joint press statement (New Zealand Government, 2016). *The New Zealand Herald* headlined its article on the plan's release 'Revealed: The bold plan to save Hauraki Gulf'. Science Reporter Jamie Morton went on to write '[t]he plan – the first of its kind in New Zealand – sought to help stem the flow of sediment and other pollutants into the Hauraki Gulf, ease pressures on wildlife, fish stocks and kaimoana and restore the health of crucial ecosystems' (Morton, 2016). Some other responses were less positive. A major fishing company expressed reservations about the proposal to phase out bottom-impacting fishing (Bradley, 2016). Another commentator criticized the plan as being undemocratic due to the adoption of a collaborative stakeholder-led process for its development (Fox, 2016).

More than a year after the plan's release, however, there has been little progress on the ground with implementation. This reflects the difficulty identified by Kidd and Ellis (2012) that, although considerable effort is often made to give plans scientific rigour and public legitimacy, much less attention is given to their implementation and there is often a gap between the content of marine plans and the ability to deliver on their ambition. The seeming reluctance of some agencies to implement a plan for the Hauraki Gulf that they did not have a strong hand in developing highlights some of the issues with delegating plan development to non-governmental groups. In addition, the current fragmented state of governance of the Hauraki Gulf also likely makes it difficult to implement an integrated plan. The SWG was well aware of this problem and the plan itself expresses a view on 'some attributes of future governance of the Hauraki Gulf Marine Park that we believe are essential for the implementation of this Plan' (Sea Change Tai Timu Tai Pari, 2017, 179). A key element of this is 'strong, effective co-governance' with a new governance entity having 'membership from Mana Whenu and the community at large' (Sea Change Tai Timu Tai Pari, 2017, 179). Fourteen functions of such a new entity are identified in the plan, including leading strategic Gulf-wide initiatives, overseeing the design of a detailed implementation plan, providing recommendations to the Minister for Primary Industries on fisheries sustainability measures and regulations applying to the Park and helping to establish the network of MPAs identified in the plan (Sea Change Tai Timu Tai Pari, 2017, 179–180). This new governance entity could take the form of a reconfigured Hauraki Gulf Forum, which would require legislative amendments to the Hauraki Gulf Marine Park Act 2000. It could potentially draw on the skill base developed within the SWG during the process.

Collaborative marine planning from the "coalface"

This chapter has, in large part, presented the 'nuts and bolts' process of attempting to form a marine spatial plan. My own experience of the collaborative planning process, however, provides a more embodied flavour of how the dynamics of the process works. This more personal dimension is vital. Arguably it is worth those engaging with MSP to consider their own experiences, which are inevitably bundled up in the more procedural processes described.

When the SWG first met, most of the members were strangers to me and I was suspicious of their motivations. I assumed that the representatives of the commercial sectors, such as fishing, farming and aquaculture, were there to protect their interests and to secure any commercial gains they could. I was unsure of the motivation of Mana Whenua members, but was concerned that Māori interests in the commercial fishing industry would outweigh an environmental ethic. Somewhat surprisingly, although I had worked in the field of marine policy for some years, I had never actually spoken to a commercial fisher and now I was sitting around the table facing two of them.

It took some months for a relationship of trust to develop between the SWG members and this strengthened considerably over the three-year period. It eventually became evident that most of those sitting around the table had a strong commitment to doing what was in the best long-term interests of the Gulf, rather than in trying to benefit their own sector in the short term. This was crucial to the success of the plan, because during the process, and particularly near its end, there were some very contentious issues to resolve. At times there was a real danger that some parties would walk away from the process, but I believe that it was this commitment built up from personal associations with the Gulf that kept people around the table.

Although much science has been carried out on the Gulf, we soon discovered that there were many significant gaps, and that those who had spent much of their lives on the water had a wealth of knowledge. This knowledge was only shared with the group later on in the process, after a strong climate of trust had been developed. As the SWG members went through a process of jointly receiving information on the Gulf, and listening to each other's views and values, their positions started to shift and converge. Reaching agreement then became possible on topics for which there had historically been high levels of conflict.

The Sea Change Tai Timu Tai Pari process has enabled me to develop valuable relationships with other sectors which I continue to benefit from. I now have good contacts in the commercial and recreational fishing sectors who I can readily ring to discuss matters and I am able to directly contact Mana Whenua members regarding broader matters. This was not possible before. I believe that the greater knowledge and stronger network of relationships that many SWG members developed through the process can assist with the ongoing implementation of the plan.

Conclusions

The Hauraki Gulf has a long history of human association which has served both to engender close human relationships with the place, but also to drive considerable environmental pressures on the marine system. The Sea Change Tai Timu Tai Pari marine spatial planning process was able to draw on these strong social and cultural relationships with the Gulf, as a common thread to draw disparate sectors together in order to reach consensus on how to address these pressures. Implementation of the plan is proving a challenge, and new ways may need to be found to extend this joint inter-sectoral commitment to the implementation phase.

This chapter has presented the challenges and opportunities in working collaboratively to address the degradation of a significant marine space. Whilst progress has been slow, the process has illustrated how, collectively, steps can be made to improve our relationship with the sea in a manner which is inclusive and forward-looking. The insights provided here may help to shape plans in other places in order to improve how we might better live with the sea for the benefit of all.

Notes

1 The sliding baseline phenomenon refers to the 'incremental lowering of ecological standards' caused by a reduction in expectations of what the environment should look like due to basing such expectations on what has been experienced during one's lifetime rather than on robust historical data (Gatti et al., 2015, 2).
2 A New Zealand environmental NGO for which I work and which had for some years promoted the adoption of MSP within the country.

References

Beverley, P. (2015). A stronger voice for Māori in natural resource governance and management. *Planning Quarterly*, 198, 13–17.

Bradley, A. (2016). Fishing Giant Sanford rails against Hauraki Gulf plan. *Radio New Zealand*. [Online] http://www.radionz.co.nz/news/national/319795/fishing-giant-rails-against-hauraki-gulf-plan (Accessed 23 January 2018).

Bradly, N. (2015). *Review of the Hauraki Gulf Forum*. Auckland: Envirostrat Consulting Limited.

Brower, A. L. (2016). Is collaboration good for the environment? Or, what's wrong with the land and water forum? *New Zealand Journal of Ecology*, 40(3), 390–397.

Constantine, R., Johnson, M., Riekkola, L., Jervis, S., Kozmian-Ledward, L., Dennis, T., Torres, L. G., and Aguilar De Soto, N. (2015). Mitigation of vessel-strike mortality of endangered bryde's whales in the Hauraki Gulf, New Zealand. *Biological Conservation*, 186, 149–157.

Department of Conservation. (2017). *Te Hauturu-o-Toi Little Barrier Island Nature Reserve 2017 Management Plan*. Wellington: Department of Conservation.

Dwyer, S. L., Tezanos-Pinto, G., Visser, I. N., Pawley, M. D. M., Meissner, A. M., Berghan, J., and Stockin, K. A. (2014). Overlooking a potential hotspot at Great Barrier Island for the nationally endangered bottlenose dolphin of New Zealand. *Endangered Species Research*, 25, 97–114.

Ehler, C., and Douvere, F. (2009). *Marine Spatial Planning: A Step-by-Step Approach Toward Ecosystem-Based Management*. Paris: UNESCO.

Fox, T. (2016). Comments by Tony Fox – TCDC Councillor for Mercury Bay. *Mercury Bay Informer*, 720, 14–16.

Furey, L. (1997). *Archaeology in the Hauraki Region: A Summary*. Paeroa: Hauraki Maori Trust Board.

Gaskin, C., and Rayner, M. J. (2013). *Seabirds of The Hauraki Gulf: Natural History, Research And Conservation*. Auckland: Hauraki Gulf Forum.

Gatti, G., Bianchi, C. N., Parravicini, V., Rovere, A., Peirano, A., Montefalcone, M., Massa, F., and Morri, C. (2015). Ecological change, sliding baselines and the importance of historical data: Lessons from combing observational and quantitative data on a temperate reef over 70 years. *PLoS One*, DOI: 0118581. https://doi.org/10.1371/journal.pone.0118581

New Zealand Government. (2016). Ministers welcome release of sea change plan. Press Release, 7 December.

Green, M., and Zeldis, J. (2015). *Firth of Thames Water Quality and Ecosystem Health: A Synthesis*. Hamilton: NIWA.

Hako, N. T. K. T., Hei, N., Maru, N., Paoa, N., Hauraki, N. P. K., Pūkenga, N., Tumutumu, N. R., Tamaterā, N., Tokanui, N. T., Whanaunga, N., and Patukirikiri, T., and The Crown. (2017). *Pare Hauraki Collective Redress Deed*. [Online] https://www.govt.nz/dmsdocu ment/7118.pdf (Accessed 23 January 2018).

Hauraki Gulf Forum. (2011a). *State of Our Gulf 2011*. Auckland: Hauraki Gulf Forum.

Hauraki Gulf Forum. (2011b). *Spatial Planning for the Gulf: An International Review of Marine Spatial Planning Initiatives and Application to The Hauraki Gulf*. Auckland: Hauraki Gulf Forum.

Hauraki Gulf Forum. (2015). *Annual Report 2014–2015*. Auckland: Hauraki Gulf Forum.

Hauraki Gulf Maritime Park Board. (1968). *Hauraki Gulf Maritime Park: Areas which have been Formally Added to the Hauraki Gulf Maritime Park as of 14 November 1968*. Auckland: Hauraki Gulf Maritime Park Board.

Holley, C., and Gunningham, N. (2011). Natural resources, new governance and legal regulation: When does collaboration work? *New Zealand Universities Law Review*, 24(3), 309–336.

Independent Review Panel. (2012). *First Review Report*. Auckland: Auckland Council.

Interagency Ocean Policy Task Force. (2009). *Interim Framework for Effective Coastal and Marine Spatial Planning*. Washington DC: The White House Council on Environmental Quality.

Jay, S., Ellis, G., and Kidd, S. (2012). Marine spatial planning: A new frontier. *Journal of Environmental Policy & Planning*, 14(1), 1–5.

Kidd, S., and Ellis, G. (2012). From the land to sea and back again? Using terrestrial planning to understand the process of marine spatial planning. *Journal of Environmental Policy & Planning*, 14(1), 49–66.

Kidd, S., and Shaw, D. (2013). Reconceptualising territoriality and spatial planning: Insights from the sea. *Planning Theory & Practice*, 14(2), 180–197.

Morrison, M. A., Jones, E., Consalvey, M., and Berkenbusch, K. (2014). Linking marine fisheries species to biogenic habitats in New Zealand: A review and synthesis of knowledge. *New Zealand Aquatic Environment and Biodiversity Report No.*, 130.

Morrison, M. A., Lowe, M. L., Parsons, D. M., Usmar, N. R., and McLeod, I. M. (2009). A review of land-based effects on coastal fisheries and supporting biodiversity in New Zealand. *New Zealand Aquatic Environment and Biodiversity Report No.*, 37.

Morton, J. (2016). Revealed: The bold plan to save Hauraki Gulf. *The New Zealand Herald*, 6 December.

O'Brien, J. D. (1971). Preservation of our national seashores. *Proceedings of the New Zealand Ecological Society*, 18, 8–12.

Paul, L. J. (1977). The commercial fishery for snapper, *Chrysophyrus auratus* (Forster) in the Auckland Region, New Zealand, from 1900 to 1971. *Fisheries Research Bulletin*, 15.

Paul, L. J. (2012). A history of the Firth of Thames dredge fishery for mussels: Use and abuse of a coastal resource. *New Zealand Aquatic Environment and Biodiversity Report*, 94.

Peart, R. (2016). *The Story of the Hauraki Gulf*. Auckland: David Bateman Limited.

Peart, R. (2017). *Sea Change Tai Timu Tai Pari Case Study*. Unpublished Report prepared for the Sustainable Seas National Science Challenge, Auckland: Environmental Defence Society.

Reed, A. H., and Reed, A. W. (Eds.). (1951). *Captain Cook in New Zealand*. Wellington: A. H. & A. W. Reed.

Reed, A. W. (2001). *Illustrated Māori Place Names*. Auckland: Reed Books.

Salmond, A. (2017). *Tears of Rangi: Experiments Across Worlds*. Auckland: Auckland University Press.

Sea Change Tai Timu Tai Pari. (2013). *Stakeholder Working Group: Terms of Reference*. Auckland: Auckland Council.

Sea Change Tai Timu Tai Pari. (2017). *Hauraki Gulf Marine Spatial Plan*. Hamilton: Waikato Regional Council.

Smith, I. W. G. (1989). Māori impact on the marine megafauna: Pre-European distributions of New Zealand marine mammals. In D. G. Sutton (Ed.), *Saying so Doesn't Make it so*, Auckland: New Zealand Archaeological Association, pp. 76–108.

Stone, R. C. J. (2001). *From Tāmaki-Makau-Rau to Auckland*. Auckland: Auckland University Press.

Waitangi Tribunal. (2001). *Hauraki Gulf Marine Park Act Report*. Wellington: Waitangi Tribunal.

Waitangi Tribunal. (2006). *The Hauraki Report*. Wellington: Waitangi Tribunal.

Waitangi Tribunal. (2011). *Waitangi Tribunal, Ko Aotearoa Tēnei: A Report into Claims Concerning New Zealand Law and Policy Affecting Māori Culture and Identity: Te Taumata Tuatahi*. Wellington: Legislation Direct.

Watton, G. (1995). *Taming the Waihou: The Story of the Waihou Valley Catchment Flood Protection and Erosion Control Scheme*. Hamilton: Waikato Regional Council.

Wright, O. (1950). Ed. *New Zealand 1826–1827 from the French of Dumont D'Urville*. Wellington: Wingfield Press.

Yaffee, S. L. (2002). Benefits of collaboration. [Online] http://www.snre.umich.edu/ecomgt/lessons/stages/getting_started/Benefits_of_Collaboration.pdf (Accessed 23 January 2018).

Statistics New Zealand. (2017). Auckland's future population under alternative migration scenarios. [Online] http://archive.stats.govt.nz/browse_for_stats/population/estimates_and_projections/auck-pop-alt-migration-2017.aspx (Accessed 23 January 2018).

Zeldis, J. R., and Francis, R. I. C. C. (1998). A daily egg production method estimate of snapper biomass in Hauraki Gulf, New Zealand. *ICES Journal of Marine Science*, 55, 522–534.

Zeldis, J. R., Walters, R. A., Greig, M. J. N., and Image, K. (2004). Circulation over the northeastern New Zealand continental slope, shelf and adjacent Hauraki Gulf, during spring and summer. *Continental Shelf Research*, 24, 543–561.

4 Geo-spatial analysis

Assessing the multiple values and complexity of seascape

Lars Brabyn

How might we better gain knowledge of people's awareness of the sea, their relationship to it, and any possible actions that may result? This chapter describes one method for assessing the 'value' of seascapes (see Burgess and Gold, 1982; Carolan, 2013), and the relationship people have with the sea and coastal areas. Arguably, alongside more phenomenological, subjective and discursive understandings of engagements with the sea, increased awareness and interactions with seascape environments also requires information that is based on methods that can be validated numerically. Information that is based on opinions, feelings and beliefs is vital but, of course, can be counteracted by opposing opinions. Policy, action and change therefore tends to rely on information that can be independently reproduced and validated.

This chapter draws on psychophysical assessment and how it can be implemented using Geographical Information Systems (GIS) in order to assess perceptions of seascapes. An example, using the landscape analysis that draws in the seascapes of the southern South Island of New Zealand, is provided to demonstrate how the method works. GIS was first used to classify the landscape character, which provided the vital frame of reference for communication. Yet, as this chapter shows, classifications and assessments of landscape character may also be used to provide a framework for understanding seascape.

Studies of 'landscape' have a long tradition in geography (Wylie, 2007). Early work concerned with landscape, for example Carl Sauer's *Morphology of Landscape* (1925), provided early attempts to describe landscape as a term and to map specific physical traits or what he called 'imprints of man' to the earth. Later, landscape content category research in the 1970s determined that people conceptualised the character of landscapes using classes of landform, landcover, natural character (conversely human development) and water. In the 1980s and 1990s work became concerned with 'visions' of landscape that were constructed as 'authoritative' representations of place (Duncan, 1995), and more recent work has sought to consider embedded landscape engagements (Stephenson, 2008). With a recent upturn in the use of spatial analysis and GIS, this technology and the techniques it enables arguably provide another lens for working with landscape.

Whilst the term landscape has a long history and usage – and indeed, this chapter will draw on landscape assessment – the term 'seascape' can be part and parcel of such an assessment. Moreover, the term has some similarities with landscape in that the character of both is predominantly perceived as visual (Cosgrove, 2008; Wylie, 2007). However visions of seascape are often restricted to the coastal region where the sea is only part of the view or experience. Like landscape, though, the seascape is a multi-layered concept that consists of the physical character of the coastal region as well as further afield, and the many different values and associations that are evoked when different people perceive these places. The physical character of the seascape can have many different permutations of landform, landcover, infrastructure and sea conditions. The coastline can be indented, such as an estuary, fiord or bay, and have relatively flat and protected sea conditions. Conversely, the seascape can consists of open sea, wild weather and tumultuous sea conditions. The seascape is the totality of the physical character and it is problematic to isolate the different physical components. For example, a mountain by the sea that is covered in forest is different to a mountain by the sea without forest, or a flat forested area by the sea. As such, this chapter brings together land- and seascapes, demonstrating how each combination of landform, landcover and infrastructure provides a different *seascape*.

Moreover, like landscape, what makes seascape complex is that it also involves perception – the perceived seascape. The seascape is not only dependent on physical characteristics but also on how people see, interpret and conceptualise what is 'out there'. For a given coastal view, two people can see two very different seascapes because their associations with the coast may be very different. One person may be an overseas tourist who has never seen this coast before, while another person may have lived on this coast all their life, as have previous generations of their family. One person may be a conservationist who wants to preserve nature, while another person is an entrepreneur looking for opportunities to profit from the land and sea. For this reason *values*, including place attachment, become an important part of seascape assessment which includes recreation, biodiversity, wilderness, economics and history. It is because of the multitude of values associated with seascapes that these places can be contentious for spatial planning and land-use management. Different industries can be at odds over appropriate forms of development. The tourism industry may want the preservation of natural character, while the forestry industry may want a port facility. Such contention can result in costly environmental court cases and discontent over both change and non-change.

Theories associated with value, sense of place and place attachment have circulated since Tuan's (1977) early work in this field. People develop associations to places that become part of their cultural identity; they have 'embedded values' based on historical connections, as opposed to 'surface values' based on just present experiences (Stephenson, 2008). Landscape values have long been recognised and can be categorised by the triple bottom line approach – environmental, social and economic – or be part of more elaborate typologies of values (see Brown and Reed, 2000; Stephenson, 2008). An important understanding of landscape, which is often overlooked, is that 'beauty is in the eye of the beholder'. For a given coastal location, every person sees a different seascape and has individual values associated with this view. Therefore, so-called seascape assessments based on the expert approach, which typically relies on form and function (Daniel and Vining, 1983), are theoretically invalid because perception is an important component of seascape.

It is argued in this chapter, then, that seascape is taken as the totality of both the physical and perceived and therefore the value of a seascape cannot be calculated by the sum of the parts that construct the physical and perceived layers of the seascape; instead the seascape needs to be considered and assessed in totality. To assess seascape (through a landscape analysis) in such a way is a daunting task, but Geographical Information Systems combined with internet technologies can offer cost-effective solutions to this challenge, and to policy-orientated work that helps validate how we live with the sea. GIS is a form of computer-based mapping that links data sets containing location information to a map graphic and includes a wide range of spatial analysis tools that replicate the manual assessment of maps. A known example of GIS interfacing with the internet is Google Maps. These maps are a powerful method for providing and sharing information about places.

The assessment of seascape values is critical for the wise management of coastal regions. The coast is a region of intense development pressure because it is a desirable place for people to live, provides for recreation and tourism, supports marine farming and provides access for transportation through ports (see also Chapter 3). To ensure that all the benefits and costs associated with different forms of development are well considered, arguably it is useful for seascapes' values to be assessed and represented. The research challenge is to develop assessment methods that capture the totality of the multiple physical and human layers that are a seascape. This departs from other more qualitative 'landscape' (Wylie, 2007) and 'seascape' (Brown and Humberstone, 2015) work, but demonstrates a novel and increasingly important way in which spatial engagements are researched, assessed and understood.

Psychophysical methods

Having provided an introduction, it is now argued that psychophysical analysis provides a valid method for landscape assessment, combining human mental processes with physical landscapes (Daniel and Vining, 1983). The psychophysical approach to assessment has until recently been expensive to implement because it requires both a landscape character classification and a means of determining landscape preferences. However, the use of GIS makes this approach less expensive and therefore more accessible. Typically, the method has involved asking people to rank landscapes using photographs as representations of landscapes. A classification of the physical landscape is used to determine what landscape each photo represents. Using statistical techniques, preferences scores for different landscape character classes are derived and these are applied to areas with these landscape classes. The psychophysical method has the advantage of incorporating the connections people have to landscapes and thereby helps to account for perception and how 'beauty is in the eye of the beholder'. However, the method is problematic because photos are always representations, and are moreover two-dimensional compared to the three-dimensional landscapes. Furthermore, it is costly to implement. As a result, rather than trying to incorporate everyday perceptions, 'expert judgements' based on form and function have dominated landscape assessment because they are expedient, even though this method is invalid.

In implementing a psychophysical method, however, there are two main components. The first is a landscape character classification that provides information on the physical landscape of a region. This classification provides a frame of reference for communication and enables survey results to be extrapolated over a larger region. Just like botanists, who utilise a plant classification to communicate what plant they are researching, a landscape classification helps communicate what type of landscape is being researched and the extent of this landscape. It is impractical and difficult to survey landscape preference for every location, so extrapolation is a common practice for the assessment of spatially distributed resources. The second main component of the psychophysical method is the surveying of the landscape preferences people have. In spite of its critiques, as demonstrated here, it can be valuable to corporate understandings of landscapes – and indeed seascapes – beyond the 'expert' *view*point.

New Zealand landscape character classification

The NZ Landscape Character Classification (NZLCC) was developed by Brabyn (1996) and subsequently updated (Brabyn, 2009).[1] The NZLC is a classification of character, not quality, and is built from the unique combinations of six landscape components: landform, landcover, infrastructure, water, dominant landcover and water views. The latter two components provide a wider experiential context of a place. The categories associated with landform and landcover are listed in the results tables (see Tables 4.1 and 4.2). As noted, although such analysis seems 'land' focused, each combination of landform, landcover and infrastructure provides a different *seascape.*

The purpose of landscape classification, then, is to provide a frame of reference for communicating landscape research to relevant agencies and other researchers and this can include communicating seascape research too. The classification system uses common language to describe the landscape components and component categories. The six landscape components have the potential to produce many thousands of landscape classes – unique combinations of components – which may be impractical for some applications. Consequently, a hierarchical structure is imposed on the classification system so users can select a level of generalisation to meet their needs. This study uses the most detailed level (called 3a). The NZLCC is operationalised and accessed as a GIS database. The advantage of a GIS-based classification system is that statistics on the total area and relative abundance of each landscape class at a regional or national level can be calculated to help interpret the results.

Using public participatory GIS (PPGIS) to survey landscape preferences

A relatively new form of internet GIS is called Public Participatory GIS (PPGIS), whereby maps on the internet have enabled the public to provide information about places. Maps on the internet can be used not only to provide information about places but also to collect information about places. PPGIS can be used to collect information on the values and associations that people have with the physical landscape. In essence PPGIS can be used to represent the multiple perceived layers of the landscape (Brown and Reed, 2000).

Figure 4.1 shows the PPGIS internet site used in this research. On the left is a range of icons representing different values and on the right is a map using a custom Google® maps application. Participants of the site can drag and drop different value icons at different locations on the map to show what sites they value. The users can zoom in to the necessary level of detail in order to accurately position the location that they want to represent. The zoom level which people use to position their value is recorded, and the site can be set up so that value icons cannot be placed when the scale is too broad. Once a participant has finished locating their value icons, the coordinates, value type and zoom level are recorded into a central database for further analysis. Participants also have the opportunity to enter comments associated with their values. The site can also request demographic information about the participant before they can place values.

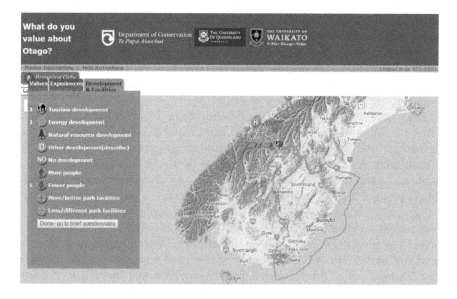

Figure 4.1 The internet interface for the Public Participatory GIS. By the author.

Access to a PPGIS website can be controlled by a password or can be made publicly available. The advantage of using a password is that people can be mailed a letter introducing the study and asking then to participate. The password is then linked to their address, and the postcode can be used to analyse the spatial variations in responses. For example, participants in Dunedin, New Zealand can be compared with participants from Queenstown, New Zealand or the distance from home of the preference sites can be studied.

An important part of the PPGIS site is the *choice* of landscape values that can be used. Landscape values are best viewed as a type of 'relationship' value that bridges held and assigned values (Brown and Weber, 2012). In the process of associating meanings with place, what is important to a person becomes fused with conceptions of what appears important to them in the physical landscape (and indeed seascape). In the PPGIS mapping process, individuals call upon their tacit, held values in the process of assigning values to landscapes such as those in southern New Zealand. In PPGIS, an operational definition for each landscape value is provided to the study participant.

The PPGIS study demonstrated in this chapter was developed after consultation and pilot testing with Department of Conservation (DOC) staff in New Zealand. Participants were recruited January to March 2011 through a random mail sample of households in the Southland and Otago regions, by visitor contact at conservation areas and by advertising in media outlets such as local newspapers. Participants were instructed to drag and drop these markers to the appropriate location on a Google® base map. The map did not contain the information present in the NZLCC but instead had shaded relief and standard topographical detail so that participants could accurately locate particular places. Of relevance to this study was the list of 11 landscape values that appear in Table 4.1, which are based on the values used by Brown and Reed (2000). These landscape values provide a broad range of values that are relevant to regional conservation and development issues in New Zealand, and arguably more broadly. For purposes of this study, the two categories of recreation value were combined into a single recreation value while the two separate values of native vegetation and native wildlife were combined into a single native flora/fauna value category. Eight landscape values (aesthetic/scenic, recreation, economic, ecological/life sustaining, native flora/fauna, social, historical/cultural, wilderness) were included in the analysis.

PPGIS mapping precision by participants was enabled by only allowing the placement of markers if the participant had zoomed in to a predetermined zoom level (Level 12) in Google® Maps (approximately 1:100,000 scale). Respondents could optionally view the region in different Google® map views including 'Map', 'Terrain', 'Satellite', 'Hybrid' and 3-D 'Earth'. Participants were instructed to drag and drop these markers to the appropriate location on a Google® base map. The map did not contain the information present in the NZLCC but instead had shaded relief and standard topographical detail so that participants could accurately locate particular places.

Table 4.1 Landscape value definitions used in public participation GIS (PPGIS) process in New Zealand. By the author.

Landscape values

Scenic/aesthetic – these areas are valuable because they contain attractive scenery including sights, smells and sounds.

Recreation (non-facility based) – these areas are valuable because they provide dispersed recreation opportunities where users are relatively self-reliant, i.e. tramping (trekking/backpacking), climbing, hunting/fishing or adventure activities.

Recreation (facility based) – these areas are valuable because they provide recreation activities through the provision of managed tracks, huts, campsites and other facilities.

Economic – these areas are valuable because they provide income and employment opportunities through industries like tourism, natural resources or other commercial activity.

Ecological/life sustaining – these areas are valuable because they help produce, preserve and renew air, soil and water.

Native wildlife – these are valuable because they provide areas for indigenous (native) wildlife to live and/or opportunities for humans to observe.

Native vegetation – these areas are valuable because they sustain areas of indigenous (native) plants.

Marine – these areas are valuable because they support marine life.

Social – these areas are valuable because they provide opportunities for social interaction.

Historical/cultural – these areas are valuable because they represent history, NZ identity or provide places where people can continue to pass down memories, wisdom, traditions or a way of life.

Wilderness – these areas are valuable because they are wild, uninhabited or relatively untouched by human activity.

Assessing seascapes values in the southern South Island by combining the NZLCC and PPGIS

Study location

As noted, the two regions in this study are the Otago and Southland regions on the South Island of New Zealand (see Figure 4.2). The Southland region covers more than 3.1 million hectares, has over 3,400km of coastline, and includes New Zealand's largest national park, Fiordland National Park. Southland is one of New Zealand's most sparely populated regions (approximately 94,200 in 2010) with an economy based in tourism, agriculture, fishing, forestry and energy resources.

The Otago region covers approximately 3.2 million hectares with an estimated population of 207,400 in 2010. Major centres of population include Dunedin, Oamaru, and the tourist centres of Queenstown and Wanaka. In the west of the region, high alpine mountains and glacial lakes dominate the landscape including Mt. Aspiring National Park. Tussock grasslands dominate the dry lands of the central region, while the hill country of the Catlins is located in the south-east of the region. Key economic sectors include tourism, education, agriculture and manufacturing.

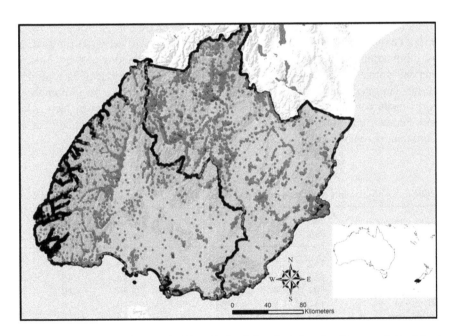

Figure 4.2 Map of study area in New Zealand that includes the Otago and Southland regions. Mapped landscape values (n≈9,000) appear in the study region as points. By the author.

Data analysis

To prepare the spatial data for analysis, eight PPGIS landscape values were intersected with the six NZLCC landscape components (landform, landcover, dominant landcover, water, water view and infrastructure) and the NZLCC landscape classes, which are combinations of the six landscape components. This produced a large table consisting of all the points collected, the landscape value associated with the point, the categories for each of the six landscape components and the landscape class.

This data was analysed using frequency counts and percentages, as well as comparisons with the land area of different landscape components. The numbers of landscape values falling within each landscape component were summed. This identified the most popular landscape components and classes associated with different landscape values (see Table 4.2). However, simple frequencies can be misleading when landscape classes are disproportionately over- or under-represented in the study region. If a landscape class is uncommon but has many landscape values, or if a landscape class is abundant but does not have high value counts, these findings merit attention. To account for proportional differences in landscape classes, area information needs to be considered.

Participant characteristics and number of landscape values

A total of 14,370 landscape attributes were identified by 608 PPGIS participants in the Otago and Southland regions. This spatial data was prepared for analysis by eliminating markers placed outside the two study regions and by filtering markers unrelated to the seven landscape values in the study, leaving a total of 10,110 landscape value points available for analysis. This is a large number of participants compared to traditional landscape perception studies in New Zealand. For example Fairweather and Swaffield's (1999) study of perceptions in the Coromandel region had only 88 participants.

Table 4.2 Value counts for coastal and non-coastal regions. By the author.

Value category	Coastal count	Coastal (%)	Inland count	Inland (%)	Land and coast count	Land and coast (%)
Flora and Fauna	1152	46	1871	25	3023	30
Recreation	536	21	3010	40	3546	35
Scenery / Aesthetic	504	20	1552	20	2056	20
Historical	137	5	383	5	520	5
Wilderness	65	3	421	6	486	5
Economic	51	2	156	2	207	2
Social	50	2	222	3	272	3
Total	2495	25	7615	75	10110	

Of the 354 participants that responded to socio-demographic questions following the PPGIS mapping activity, 94% were New Zealand residents and 6% were international visitors. The New Zealand participants were 62% male, had a median age of 48 years, with 41% reporting a bachelor's degree or higher in formal education. The, PPGIS participants were more often male, older and have more formal education than the general NZ population. These results are consistent with other internet PPGIS studies reporting higher participation by older males with more formal education (Brown, Montag, and Lyons, 2012). This bias is a limitation of the results, which future studies could improve by promoting the study to a wider group of participants.

PPGIS participants were also disproportionately represented by individuals with a self-reported 'good' or 'excellent' knowledge of places in the region (68%), compared to 'average' (26%) or 'below average' knowledge (5%). Thus, PPGIS participants report a relatively high level of familiarity with the regional landscape, a finding consistent with other PPGIS studies (see, for example, Brown and Weber, 2012; Brown, Montag, and Lyon, 2012; Brown, 2005).

Relative frequency of seascape values

Table 4.2 provides a count and percentage breakdown of the number of points collected for each landscape value, and compares coastal areas with land areas, and both combined. Of the 10,110 value points collected for both coastal and inland regions, 2,495 (25%) were in coastal regions, which consists of 1,186 value points placed in the sea, and 1,309 on land with coastal views. The proportion of the total values placed on land with coastal views is 12.9%, which is high given that this land is 1.7% of the total land area of the study region. This substantiates the importance of seascapes, which is already well understood in New Zealand (Fairweather and Swaffield, 1999).

What has not been substantiated in previous studies is the relative proportion of *values* associated with seascapes. There are three 'landscape' values that dominate how seascapes are valued: recreation, aesthetic and native flora/fauna. Interestingly, flora and fauna values are proportionally higher in coastal regions compared to inland regions, with scores of 46% and 25% respectively. For both coastal and inland regions recreational values are proportionally similar. Interestingly, though, historical values are relatively low in count, yet the study area has many important historical sites (for example, Bannockburn – see Stephenson, 2008). The results indicate that historical/cultural landscape values may not be as important as some researchers would suggest (Stephenson, 2008). The low number of wilderness values can be explained by the relatively low number of people who actively engage in wilderness activity, even though the study area contains some of the largest wilderness areas in NZ – Fiordland and Aspiring National Parks.

Moving on, Tables 4.3 and 4.4 show the types of landform and landcover. Previous studies have shown that people have a preference for natural as well as mountainous landscapes, but the results from this study show that in *coastal*

Table 4.3 Frequency of landform classes for coastal regions. By the author.

Landform	Just coast count
Sea	1,186
Mostly flat	175
Low plateau	20
Low hill	236
Hill	264
Plateau	1
Open valley with hill	63
High hill	380
Mountain	120
Open valley with mountain	9
High mountain	9
Very high mountain	32

Table 4.4 Frequency of landcover classes for coastal regions. By the author.

Landcover	Just coast count
Urban	143
Mine or dump	3
Exotic scrub	24
Exotic forest	55
Low producing grassland	73
High producing grassland	404
Indigenous scrub	145
forest	323
Tussock	6
Freshwater wetland	19
River	13
Saltwater wetland	82
Coastal sand	19
Sea	1,186

regions a range of more gentle topographies dominate even though a large proportion of the study area contains Fiordland National Park, which is world renowned for its mountainous fiords. Of the value points on land, 13% are in mountainous topography and 87% are in hilly to flat topography. This is the reverse of what is expected based on previous studies. The same reverse trend can be found with landcover. Previous studies show a preference for natural areas. Of the 1,309 value points placed on land with coastal views, 607 (46%) are found in natural or indigenous landcovers and 702 (54%) are found in unnatural landcovers, such as farmland and urban places.

Discussion

This chapter began with the following question: 'how might we better gain knowledge of people's awareness of the sea, their relationship to it, and any possible actions that may result?' This chapter has subsequently described one method for doing so, landscape research that extends to assess the 'value' of seascapes and the relationship people have with the sea and coastal areas. Seascape is a known attraction and this study confirms the importance of coastal views in landscape research. Seascapes are valued in a range of landcover and landform contexts, even in agricultural areas with low topography. Indeed, as this research has explained, each combination of landform, landcover and infrastructure provides a different *seascape* which can then be assessed.

Generally in landscape research, natural views are valued but it appears that with seascape both natural *and* developed settings are valued. This finding has not been substantiated in previous studies in New Zealand, or elsewhere, but should not appear surprising. People often enjoy 'vernacular' landscapes, perhaps because they are common and easily accessible. This can be explained by the theory of spatial or place-based discounting (Norton and Hannon, 1997; Perrings and Hannon, 2001). According to this theory, humans tend to discount across both time and space, placing higher value on places that are more proximate. In testing this theory, Brown, Reed and Harris (2002) found landscape values to be unevenly distributed across a region in Alaska, with significant spatial clustering of values near communities. Similarly, in this study, seascape values also spatially clustered in and near urban areas in the region (for example, Dunedin and Invercargill). There are many value points around the Otago Harbour, near the city of Dunedin, even though this is a relatively small proportion of the study area. This harbour is an important recreation and aesthetic site for the local population and many people will have close connections to this seascape.

The combination of the NZLCC for representing the physical seascape and PPGIS for assessing the perceived seascape values provides a cost-effective solution in implementing psychophysical assessment of the value of seascapes, albeit within the broader field of landscape analysis. The scale of this study exceeds other landscape perception studies in NZ. The combination of PPGIS data with a landscape classification system provides a powerful method for landscape and, by association, seascape assessment compared to traditional psychophysical assessments that use photos. The use of classification systems is common in the natural sciences because these systems provide an important frame of reference for communication, and yet their adoption in the social sciences has lagged. Without the NZLCC, detailed examination of the relationships between seascape values and physical landscapes would not have been possible. Similar classifications such as the Recreation Opportunity Spectrum have shown their worth over time (Kliskey and Kearsley, 1993; Joyce and Sutton, 2009). The use of GIS for seascape character classification and public participation in evaluation can produce substantiated information that has the potential to promote awareness and hopefully positive environmental action to protect valued seascapes.

Note

1 Brabyn's website (2012) provides details and graphical displays of this classification: http://www.waikato.ac.nz/fass/about/staff/larsb/landscape2

References

Brabyn, L. (1996). Landscape classification using GIS and national digital databases. *Landscape Research*, 21, 277–300.

Brabyn, L. (2009). Classifying landscape character. *Landscape Research*, 34, 299–321.

Brown, G. (2005). Mapping spatial attributes in survey research for natural resource management: Methods and applications. *Society and Natural Resources*, 18, 1–23.

Brown, G., Montag, J., and Lyon, K. (2012). Public participation GIS: A method for identifying ecosystem services. *Society and Natural Resources*, 25(7), 633–651.

Brown, G., and Reed, P. (2000). Validation of a forest values typology for use in national forest planning. *Forest Science*, 46, 240–247.

Brown, G., Reed, P., and Harris, C. C. (2002). Testing a place-based theory for environmental evaluation: An Alaska case study. *Applied Geography*, 22, 49–77.

Brown, G., and Weber, D. (2011). Public participation GIS: A new method for national park planning. *Landscape and Urban Planning*, 102, 1–15.

Brown, G., and Weber, D. (2012). Measuring change in place values using public participation GIS (PPGIS). *Applied Geography*, 34, 316–324.

Brown, M., and Humberstone, B. (Eds.). (2015) *Seascapes: Shaped by the Sea*. Farnham: Ashgate.

Burgess, J., and Gold, J. (Eds.). (1982). *Valued Environments*. London: George Allen & Unwin.

Carolan, M. (2013). Doing and enacting economies of value: Thinking through the assemblage'. *New Zealand Geographer*, 69(3), 176–179.

Cosgrove, D. (2008). *Geography and Vision: Seeing, Imagining and Representing the World*. London: IB Tauris.

Daniel, T. C., and Vining, J. (1983). Methodological issues in the assessment of landscape quality. In I. Altman and J. F. Wohlwill (Eds.), *Behaviour and Natural Environments*. New York: Plenum Press. pp. 39–84.

Duncan, J. (1995). Landscape geography, 1993–94. *Progress in Human Geography*, 19(3), 414–422.

Fairweather, J. R., and Swaffield, S. R. (1999). *Public Perceptions of Natural and Modified Landscapes of the Coromandel Peninsula, New Zealand* Agribusiness and Economics Research Unit. Research Report No. 241. Canterbury, New Zealand: Lincoln University.

Joyce, K. E., and Sutton, S. (2009). A method for automatic generation of the Recreation Opportunity Spectrum in New Zealand. *Applied Geography*, 29(3), 409–418.

Kliskey, A. D., and Kearsley, G. (1993). Mapping multiple perceptions of wilderness in Southern New Zealand. *Applied Geography*, 13(3), 203–223.

Norton, B. G., and Hannon, B. (1997). Environmental values: A place-based theory. *Environmental Ethics*, 19, 227–245.

Perrings, C., and Hannon, B. (2001). An introduction to spatial discounting. *Journal of Regional Science*, 41, 23–38.

Sauer, C. O. (1925). The morphology of landscape. *University of California Publications in Geography.*2(2), 19–54.

Stephenson, J. (2008). The cultural values model: An integrated approach to values in landscapes. *Landscape and Urban Planning*, 84, 127–139.

Tuan, Y. F. (1977). *Space and Place: The Perspective of Experience*. Minneapolis: University of Minnesota Press.

Wylie, J. (2007). *Landscape*. London: Routledge.

5 Education and learning

Developing action competence: living sustainably with the sea

Chris Eames

Living in an island nation, New Zealanders are never very far from the sea. Made up of two main islands and numerous smaller ones bounded by the Tasman, Pacific Oceans and Southern Oceans, Aotearoa New Zealand has a coastline of 14,000km. The long, slender shape of its main islands means the distance to the ocean is never more than a couple of hundred kilometres. The history of the landmass includes several inundations by the sea followed by orogeny (uplifting) of the land through the mechanisms of plate tectonics. This geological foundation has gifted the land a bedrock of mudstone, sandstone and limestone, creating an intimate connection between the land and the sea (Fleet, 1986). Plate boundary disturbance has also caused volcanic activity that has shaped the land and the coastal zones, and a broad continental shelf extends outside some distance from the land, influencing ocean dynamics and marine habitats. Glacial and fluvial activities have gouged deep and broad valleys at the margins between the land and the sea, creating the fiords in the south and, further north, a plethora of shallow bays and estuaries that provide safe harbours and food baskets (King, 2003).

It is hardly surprising, then, that the people of New Zealand have built an identity that is deeply connected with the sea. From the time of the arrival of the Māori people from their presumed ancestral homes of the northern Pacific (thought to be around AD 1200), humans have gathered at the coasts to make their homes and gather their food (King, 2003). Early Māori paddled their ocean-going waka (canoes) into the harbours and settled as iwi (tribes) around estuaries where kaimoana (seafood) was plentiful. Through their tikanga and kawa (cultural practices) they learned how to sustain these food supplies, using tools such rahui (collection restrictions) to maintain the populations of seafood species. This development over time of indigenous knowledge was important for a sustainable existence (Parsons et al., 2017).

The arrival of the first European settlers in sailing ships created a rather different relationship between people of this land and the sea (Salmond, 2017). Early settlers included whalers and sealers, who not only saw the marine environment as a food basket, but also a place to profit from. Over the period of 1790–1830, thousands of marine mammals were slaughtered for financial gain, to the point where the numbers of animals dwindled to uneconomic levels

(Ara, 2016), and farming became land-based as the driver of commercial enterprise.

Significant urban centres grew up around ports located on major harbours, servicing the export-led industries emerging from the developing agricultural sector. As these centres evolved, so too did the relationship between the people and the sea. A developing economy gave opportunities to move away from time-consuming subsistence living and created time for interactions with the marine environment that were not based primarily on essential food gathering for either direct use or commercial gain. People developed a recreational relationship with the sea that saw increasing numbers visiting beaches, fishing for pleasure and voyaging on the water in sailing boats. Demand for these activities saw development of coastal towns and villages, where baches (holiday homes) and camping grounds were opened for summer vacations (Ara, 2017).

For many European New Zealanders, summer meant going to the beach. A traditional shutdown of schools and businesses in January allowed families to spend long periods engaging with the marine environment, having recreational experiences that became associated with a developing culture of seascape recreation (Lloyd-Jenkins, 2016). These significant life experiences are recognised as being important in developing an environmental identity (Thomashow, 1995) and an ethic of care (Chawla and Cushing, 2007) for natural environments such as oceans, leading to what is sometimes called ocean literacy (Plankis and Marrero, 2010). However, in recent years getting to the beach and spending time in the marine environment has become more difficult for many New Zealanders. Costs associated with marine-based activities, and the amount of time available for recreation, have reduced opportunities for these enduring experiences.[1] At the same time, human impacts on the marine environment have been causing a decline in environmental health (Sea Change, 2017).

Accordingly, many New Zealanders have an environmental identity that incorporates experiences with the sea. This identity, though, is often based on values developed through their significant life experiences (Stapleton, 2015) with the sea, leading many to adopt sustainable ocean and coastal behaviours such as fishing within set limits, cleaning up beaches (Sustainable Coastlines, 2017) and protecting sand dunes. These behaviours have been supported by non-formal and informal education associated with structures such as marine reserves and quota management systems. These structures lie on a spectrum that spans a focus from education to regulation, and emphasise to varying degrees the provision of knowledge, the development of values and support for action. However, it is clear today that these structures have not prevented the development of human impact challenges for the marine environment.

In this chapter I am concerned with these challenges and the thinking and acting that has led to unsustainable behaviour in relation to the sea. I would argue that much of this type of behaviour emerges in situations where a linear and unproblematic educational process is assumed (Kollmuss and Agyeman, 2002). This is most often observed as the 'give them knowledge and they will behave appropriately' theory. A further issue, with particular relevance to marine eco-tourism, is the observed lack of 'follow-through' in sustainable action after marine experiences, despite espoused intentions from participants during the experience (Ballantyne and Packer, 2011).

In what follows, this chapter explores a different educational approach for how we might learn to live with the sea sustainably, considering an action competence model for marine education that synthesises knowledge, experience, reflection, vision for the future, values and action. While the focus is on New Zealand examples, the principles are applicable to all countries of the world which are influenced by, interact with and draw resources from the ocean environment. In short, that means every country.

Human impacts on the marine environment

Impacts on Aotearoa New Zealand's marine environment are both direct and indirect. In the former case, seafood harvesting and marine pollution are the most telling, and in the latter case, sedimentation of estuaries, freshwater pollution and coastal erosion and habitat change are significant. Commercial over-fishing in the ocean around Aotearoa New Zealand, for example, has led to reduction in populations and changes to marine ecosystems. Significant examples include: the collapse of the orange roughy fishery (Ministry of Primary Industries, 2009b) due to demand for the taste of the fish and the species' own ecology (low numbers living at great depth, long life and slow reproduction); and reduction in snapper fish numbers at inshore zones causing an over-supply of one the snapper's favourite foods, the kina (sea urchins), which in turn have over-grazed marine algae (Friends of Taputeranga Marine Reserve, 2015). Marine pollution has been caused by discharge of poorly treated sewage (New Zealand Herald, 2016), both by recreational boat owners and urban centres, and by chemical spills such as oil. The largest oil spill in a marine environment in Aotearoa New Zealand waters occurred in 2011 when the merchant vessel *Rena* went aground on a reef in the Bay of Plenty. Up to 350 tonnes of oil were estimated to have spilled into the Bay, washing up on beaches and fouling marine life such as seabirds (Science Media Centre, 2011). Land-use changes have seen the clearance of hill-slopes of stabilising native vegetation for establishment of pastoral and exotic forest industries, which has led to the severe erosion of topsoil into river catchments. This sediment load has made its way into many estuarine environments, silting them up. This has led to loss of water clarity, alteration of estuarine habitats, blocking of recreational waterways, and invasion of species such as mangroves, which have further changed the ecological and amenity values within these zones. This situation has often been exacerbated by run-off of nutrients such as nitrates and phosphates from agricultural lands (Waikato Regional Council, 2016). Lastly, coastal land-use change through development has had impacts on dune systems and littoral ecosystems, leading to erosion and disruption of seabird breeding zones.

In the future, it is likely that human-induced climate change will further impact marine environments. This is likely to manifest through warming oceans, with as yet unclear consequences for marine ecosystems but probable impacts on marine organisms with shells, and the flow on effects of sea-level rise and coastal storms (Parliamentary Commissioner for the Environment, 2015). Whilst in this chapter I am concerned with Aotearoa New Zealand, clearly these issues also play out globally.

Addressing the impacts

It is clear that previous human impacts on our marine environment are unsustainable. Whilst behaviour change is required to address these issues, a tension can be seen between the means to achieve this. On the one hand, the means can be deterministic, imposing change on individuals through directive mechanisms to achieve predetermined aims. On the other hand, the means can be emancipatory, empowering individuals with knowledge and values to achieve self-determined aims (Räthzel and Uzzell, 2009). Accordingly, there is a tension between regulation and education. These two approaches are not mutually exclusive but have different emphases and potentially different outcomes. Regulation can bring about a behaviour change in the short term, and if it becomes educative, it can lead to a more durable change. It can be relatively quick and easy to implement, but requires monitoring and may not be transferable from one domain to another (for example, someone limiting their fish catch by regulation may not transfer any thinking behind this action to not littering the beach). Education, on the other hand, tends to take longer to bring about behaviour change, but this change is likely to be long-lasting, not require monitoring and be transferable between domains (Jensen and Schnack, 1997). Both of these approaches have been used in Aotearoa New Zealand.

Two new regulatory systems and the promotion of a traditional system have been established to mitigate the effects of over-harvesting of seafood in Aotearoa New Zealand. Firstly, a quota management system was put in place for commercial fishing by the Ministry of Fisheries in 1986 (Ministry of Primary Industries, 2009a). This system is based on estimates of fish populations and imposes annual catch limits for several fish species. Quotas are divided up amongst industry operators in an attempt to ensure the long-term sustainability of fish stocks. This system is theoretically sound and may have worked for some species, although data around fish stocks are not robust, and seabird and marine mammal by-catch remain issues not clearly addressed (Forest and Bird, 2017). Secondly, in 1977 the first marine reserve was established at Goat Island, north of Auckland. Designed as a reserve for marine life in which the taking or harming of marine organisms is not allowed, it has proven to be highly successful in restoring the integrity of the marine environment and attracts large numbers of visitors each year to experience the diverse and plentiful marine life, to a certain extent causing other problems from visitor pressures (New Zealand Geographic, 1989). The concept has since been extended to a further 41 reserves in New Zealand waters (Department of Conservation, 2016), with the latest gazetted in 2015. Marine reserves now cover approximately 7% of New Zealand's coastline

(Forest and Bird, 2016). At many of these reserves, informal learning through interpretation signs is bridging the gap from regulation to education. Finally, the Māori concept of rahui is still being used to regulate the harvest of shellfish and fish species by recreational and traditional users. Māori have been awarded customary rights to take kaimoana for festivals and other occasions, but at other times strict limits, and occasionally bans, are placed on taking of particular seafood species to allow stocks to be maintained. Whilst education is likely to be provided within this tikanga it is uncertain how educative this system is within other aspects of our interaction with the marine environment.

Marine environmental education

The provision of environmental education (EE), also known as education for sustainable development (ESD) or education for sustainability (EfS), has been promoted globally and in New Zealand for some time (Parliamentary Commissioner for the Environment, 2004; UNCED, 1992; UNCSD, 2012; UNESCO, 1975, 1978) and is crucial at all age levels in addressing environmental and sustainability issues. EE is seen to encompass developing knowledge of environmental and sustainability issues, considering attitudes and values that can address these issues and taking action to resolve them. Early approaches to environmental education suggested that developing knowledge would lead to environmentally friendly attitudes, which would in turn lead to action-taking (Hungerford et al., 1980; Hungerford and Volk, 1990). This linear thinking has since been challenged and argued to be simplistic and often inaccurate (Heimlich and Ardoin, 2008; Kollmuss and Agyeman, 2002). More recent models suggest that knowledge, attitudes and behaviour are linked together in a complex structure (Capra, 2005). In addition, these structures are generally domain-specific, such that an individual who may possess certain knowledge and/or attitudes may act in a certain way towards one environmental issue, and in a completely different way towards another environmental issue (Alcock et al., 2017; Paço and Lavrador, 2017).

Recognising this complexity does not negate the importance of either knowledge or attitude development, as they are essential components of effective environmental action-taking. Indeed, as illustrated below, much research has been, and still is, predicated on exploring and developing learners' knowledge and/or attitudes towards the marine environment. For a numbers of years, researchers have been exploring knowledge of the marine environment, for example highlighting how children's understandings can be addressed in teaching and learning (Brody, 1996) and how their conceptions can help design marine aquaria exhibits (Ballantyne, 2004). Other researchers have included attitudinal development, assessing this in school- and university-level students after they have had interactions with dolphins (Barney et al., 2005), and with adults in interactions with other marine wildlife (Zeppel and Muloin, 2008).

However, researchers continue to explore and espouse the importance of public knowledge of marine issues, without any explicit acknowledgement of how that increased awareness might lead to action (behaviour change) (Fletcher et al., 2009). While others, such as Lu and Liu (2015), have explored the use of augmented reality technology to enhance learning about marine environments – which showed some real promise for children unable to gain regular and mean-ingful direct experiences of those environments – there has been little linking this to actual action-taking on behalf of those environments.

Evidence of the disconnect between knowledge and attitudes and subsequent environment-related behaviour or action in relation to the marine environment has been gleaned from research into the impacts of marine wildlife or eco-tourism experiences. This research has shown that, on the one hand, visitors to these experiences have been shown to exhibit immediate behaviour changes and to commit to some longer-term intentions to engage in marine conservation action (Zeppel, 2008). In another study, a correlation was found between knowl-edge and scuba divers and their reported environmental behaviours (Thapa et al., 2005). Additionally, researchers have found that whilst these experiences can stimulate actions, the practice of these actions tends to decline over time (Hungerford and Volk, 1990), eventually not being practised at all (Ballantyne and Packer, 2005). Unless the visitors can see a strong benefit for the continua-tion of their actions (McKenzie-Mohr and Smith, 1999), they are unlikely to sustain them unless their experiences and commitment to act are reinforced by further experiences or by other reinforcements such as through social media messages (Ballantyne and Packer, 2011). This recognition that reinforcement of some kind may be necessary to foster enduring competence in environmental action leads to a consideration of an action competence model for marine education and learning. The chapter discusses this next.

Action competence through marine education

The notion of action competence in relation to environmental education was first posed in the 1990s by researchers in the Royal Danish School of Educational Studies. Jensen and Schnack (1997) defined action competence as the ability to act, in this case with reference to the environment. They argued that 'the aim of environmental education is to make students capable of acting on a societal as well as a personal level' (Jensen and Schnack, 1997, 164). In order to do this, students need to study the root causes of environmental problems within the context of their society (Wals, 1994).

Jensen and Schnack (1997) further argued that education is not about simple behaviour modification without understanding, but about creating a democratic process of participation in which students decide for themselves the action they will take. In this paradigm, actions are considered to be consciously taken and targeted, since they are intentions based on experiences. In this sense, action is seen as more than merely behaviour change in a predetermined direction but that which requires a critical, wilful decision to act in a particular way (Breiting and Mogensen, 1999; Courtney-Hall and Rogers, 2002; Uzzell, 1999).

Equally, action is seen as different from activity, in which students undertake environmental tasks that do not involve solutions to the underlying environmental problem (an example of an activity would be litter collection on a beach, whereas an action would be addressing how to prevent littering in the first place). This distinction between action and activity is supported by both authenticity and a social scientific framework for developing knowledge and understanding of an issue (Uzzell, 1999, 404). The authenticity principle ensures actions are taken within the context of their locality (for example, their home, school, community), and that students seek to engage with the life of the local community as closely as possible. Conversely, engaging in scenarios and 'what if'-type situations would be activities only. In order for students to fully understand the issue or problem, and how it is structured within society and not the environment, Uzzell (1999) argues that students must be given scientific, social, cultural, economic and political information and experiences from which to examine the issue. He claims that too often students are given only the scientific information which does not enable them to do the societal analysis and thus fully understand the causes of the problem, as described by Jensen and Schnack (1997).

Jensen would agree with this summation of how issues need to be examined and has advocated for an interdisciplinary approach that draws on 'four dimensions of environment-related knowledge' (2002, 330). These four dimensions are: effects, root causes, strategies for change, and alternatives and visions. Action competence, then, is a process in which students identify environmental issues, determine solutions and take actions in ways that develop their competence for future action to solve or avoid environmental problems.

Jensen and Schnack (1997) further note that actions could be direct or indirect. Direct actions (people and environment relations) contribute directly to solving environmental problems, whereas indirect actions (people-to-people relations) are those which seek to influence others to contribute to solving problems. Both Jensen (2004) and Uzzell (1999) are careful to emphasise, however, that any action taken should be placed in the context of the problem to be solved. They noted in their work with school students that actions are often taken at the individual, class or even school level, but unless students are made aware of the greater problem their action is helping to solve (for example, being careful about what you tip down drains can safeguard against pollutants harming the marine environment), education may be limited. It is important then that people not only take action, but also understand why they are taking that action (Palmer, 1995).

In elucidating their concept of action competence, Jensen and Schnack (1997) identified four characteristics for learners: a knowledge and insight of the environmental problem; commitment to solve the problem; a vision for the future without the problem; and action experiences to draw upon.

A further component noted by Breiting and Mogensen (1999) is student confidence in their ability to influence environmental outcomes. In my own work with New Zealand schools, I and others have further developed these aspects to include: knowledge and understanding for decision-making; experiences; critical thinking and reflection; visioning; connectedness through values and attitudes; and planning and taking action (Eames et al., 2006, 2009). By developing these aspects through education, a learner was seen to develop action competence with respect to a particular environmental issue. Two features remained important in this concept: being action competent was not an end state and competence should continue to evolve; and action competence in one issue or domain would not presuppose action competence in another – in other words, it was domain-specific.

So, if this concept is applied to marine education, what might it look like? Take, for example, the issue of marine reserves, one pertinent to the Aotearoa New Zealand context. A putative scenario (already played out several times in reality) is that a local environmental group has suggested the creation of a marine reserve in its region. The marine reserve would aim to protect the marine ecosystem by banning all harvesting of kaimoana and establishing some informal education opportunities by encouraging marine experiences such as snorkelling, diving and glass-bottom boat tours, and installing some interpretation panels. The local community is divided on the issue: some support the reserve to create perpetuity of species populations within the reserves, maintaining biodiversity and acting as a nursery for the surrounding marine zones, and providing educational and tourism opportunities. Other locals do not support the reserve, citing their cultural heritage of taking kaimoana in that place and voicing concerns about tourism influx into their backyard.

An action competence approach within formal education settings such as schools would dictate that a teacher would provide their students with opportunities to develop knowledge through inquiry about the area – what types of organisms live there, how do they interact with each other, how have people interacted with the ecosystem (socially, culturally and economically). This knowledge development would be supported by experiences (wherever possible directly) with the environment, such as with a group like Experiencing Marine Reserves (EMR, 2016). Students could gain experiences on nearby marine reserves and contrast those with areas that are not reserved. They could augment their learning with interviews with local residents who discuss their experiences in reserved and non-reserved areas to gain wider perspectives on the issue. The students and teacher may then reflect on their learning and consider the multiple viewpoints, taking care to give weight to each, and to consider the emotional connections that these viewpoints may embody. This may lead to a visioning exercise in which the students consider evidence from established marine reserves and juxtapose this with their own experiences, reflections on the knowledge they have gained and the perspectives they have heard to consider their vision for the area concerned. Depending on the outcomes of this visioning, students may decide that something needs to be done. Educational theory at this point indicates that most successful learning and action-taking occurs when students decide for themselves (with teacher guidance) what action they should or should not take (Jensen and Schnack, 1997; Wals and Dillon, 2013). This empowers students and helps them to develop the skills for future decision-making and action-taking. Critical reflection is key to effective action choices, as assumptions embedded in social norms are challenged and alternative ways of behaving are considered. Should an action or actions be decided upon, the teacher's role then becomes important in scaffolding the students through the skills needed to plan and to take action. This may be to engage in a direct action such as planting alongside waterways that drain into the marine area to prevent sediment and nutrient run-off during rain events, or it may be more indirect by creating a pamphlet for distribution to local households, putting on a play about marine reserves for the local community, or writing to the local newspaper in support of the reserve. Of course, it is also possible that the students would decide not to support the reserve, in which case the actions would reflect that approach. Whatever action may be decided upon, it should be manageable and achievable by the age group of the students, such that the chances of successful action are enhanced (Eames et al., 2006), leading to a belief that action-taking for the environment is possible.

Designing an action competence approach for non-formal education settings where learners may attend an information day/evening hosted by educators can follow similar lines to that which happens in formal education, but learners are often adult or a mix of adult and children and therefore planning needs to take this account. Engaging in the same processes is important but there are obvious constraints, such as time. Social media has the potential to overcome some of these constraints and extend the time available for the educational process beyond the face to face component (Warner et al., 2014). Today's mobile learning technologies (e.g. smartphones, tablets) have multiple potential positive impacts for learning. They allow for learning processes to occur practically anywhere in collaboration with anyone (Cochrane et al., 2013; Pachler et al., 2010). They also promote innovative (Parsons, 2013), inclusive (Traxler, 2010) and transformative (Lindsay, 2015) types of learning that challenge traditional pedagogical approaches (Merchant, 2012). The content can be shaped to fit individual characteristics and needs (Aguayo, 2016), through self-determined and real-life learning, and within user/learner-generated content and contexts (for example *heutagogy*) (Hase and Kenyon, 2013). This wide-ranging flexibility can then be applied to (re-)connect people to marine environments and promote locally meaningful transformative learning, ecological understanding and action-taking for sustainability. These approaches can also act to create continuity between experience and later reflection.

Community-based education more often occurs in an informal context, also called free-choice learning, in which all learners (children and adults) access education when and how they wish with no direct instruction from an educator. This type of education is represented often at marine reserves by interpretation panels, which tend to focus most heavily on knowledge transmission. As noted above, whilst knowledge is a key component of developing action competence, and in informal learning contexts providing information at a site is often the most convenient, this one-dimensional focus on education relies heavily on learner motivation and insight as to what they should do with the knowledge they gain. Equally, combining the provision of knowledge through these panels and creating opportunities for experiences in, for example marine reserves, can be a powerful motivator for learners to state an intention to act. However, as discussed above, these stated intentions do not necessarily translate into actual action without some post-visit reinforcement (Ballantyne and Packer, 2011). More consideration of how the concept of action competence could better inform these educational opportunities would be valuable in the design phase. The possibility exists to design interpretation that leads visitors through the steps of developing action competence, through some knowledge provision as they reach the reserve, facilitating quality experiences in the reserve, and scaffolding opportunities to reflect, create visions and plan and take actions. Use of digital devices such as mobile technologies provide excellent potential to reinforce learning and connect to resources beyond the static display (Cochrane, 2011; Somekh, 2007).

Empowering learners in any setting through development of action competence gives them the skills to gain knowledge, recognise attitudes, and plan and take action on an issue of concern to them. These are lifelong skills that can be transferred from one domain to another to enhance sustainable living through many spheres of life.

Conclusion

Many New Zealanders strongly identify with the sea through their lifelong experiences connected with living close by or regularly visiting. Modern lifestyles and expanding populations are now threatening the breadth and depth of these experiences within society, tempering the intensity of connections to the sea. At the same moment, notions of the sea as a vast entity that could absorb any effects of human activity are being challenged by increasing understanding of human impacts upon it. Issues of declining fish stocks and increasing pollution are symptoms of societal behaviour that is affecting the marine environment.

Many other nations of the world face similar issues. Those countries that border the sea not only depend upon it for trade and sustenance, climate influence and tourism/recreation, they are also at risk from it through sea-level rise and storms as a result of climate change. Even those countries not directly connected with the ocean are not immune from these threats. This widespread interdependence between human beings and the sea and the challenges this relationship now faces suggests that our current behaviour needs to change.

Whilst regulation (with political will) can bring about fast changes to this behaviour, *education* is a more potent, enduring and wide-ranging response to these challenges and provides a pathway to effective action to (re-)create a sustainable existence for our seas. In this chapter, I have argued that the action competence concept can be applied to this goal in tackling marine environmental issues through formal, non-formal and informal education. Adopting this conceptual base can assist educative processes to move beyond knowledge transmission to provide experiences that reinforce attitudes and empower learners to take informed and intentional action to look after the sea. Knowledge and awareness remain important foundations to taking action, but we know that having knowledge is not enough to get people to act. Provision of experiences must (re-)connect people with the sea, if not physically then virtually, such that an emotional response motivates intentional action that is reinforced and supported to persist. If we can do that, we can learn to live with the sea in sustainable ways for generations to come.

Note

1 Likewise, this might not only be the case in New Zealand, but globally with the price of marine- and maritime-related activities and 'kit'.

References

Aguayo, C. (2016). Activity theory and community education for sustainability: When systems meet reality. In D. Gedera and J. Williams (Eds.), *Activity Theory in Education: Research and Practice*. Rotterdam: Sense. pp. 139–151.

Alcock, I., White, M. P., Taylor, T., Coldwell, D. F., Gribble, M. O., Evans, K. L., and Fleming, L. E. (2017). 'Green' on the ground but not in the air: Pro-environmental attitudes are related to household behaviours but not discretionary air travel. *Global Environmental Change*, 42, 136–147.

Ara, T. (2016). *Te Ara – the Encyclopedia of New Zealand* [Online] http://www.teara.govt.nz/en (Accessed 21 October 2016)

Ara, T. (2017). *Te Ara – the Encyclopedia of New Zealand* [Online] https://teara.govt.nz/en/camping (Accessed 21 October 2017)

Ballantyne, R. (2004). Young students' conceptions of the marine environment and their role in the development of aquaria exhibits. *Geo Journal*, 60, 159–163.

Ballantyne, R., and Packer, J. (2005). Promoting environmentally sustainable attitudes and behaviour through free-choice learning experiences: What is the state of the game? *Environmental Education Research*, 11(3), 281–295.

Ballantyne, R., and Packer, J. (2011). Using tourism free-choice learning experiences to promote environmentally sustainable behaviour: The role of post-visit 'action resources'. *Environmental Education Research*, 17(2), 201–215.

Barney, E. C., Mintzes, J. J., and Yen, C. (2005). Assessing knowledge, attitudes, and behavior toward charismatic megafauna: The case of dolphins. *The Journal of Environmental Education*, 36(2), 41–55.

Breiting, S., and Mogensen, F. (1999). Action competence and environmental education. *Cambridge Journal of Education*, 29(3), 349–353.

Brody, M. (1996). An assessmnet of 4th, 8th and 11th grade students' environmental science knowledge related to Oregon's marine resources. *The Journal of Environmental Education*, 27(3), 21–27.

Capra, F. (2005). Speaking nature's language: Principles for sustainability. In M. Stone and Z. Barlow (Eds.), *Ecological Literacy: Educating our Children for a Sustainable World*. San Francisco: Sierra Club Books.

Chawla, L., and Cushing, D. (2007). Education for strategic environmental behaviour. *Environmental Education Research*, 13(4), 437–452.

Cochrane, T. (2011). mLearning: Why? What? Where? How? Paper presented at the Changing demands, changing directions, Ascilite 2011, Hobart, Australia.

Cochrane, T., Buchem, I., Camacho, M., Cronin, C., Gordon, A., and Keegan, H. (2013). Building global learning communities. *Research in Learning Technology*, 21, 21955. 10.3402/rlt.v21i0.21955

Courtney-Hall, P., and Rogers, L. (2002). Gaps in mind: Problems in environmental knowledge and behaviour modelling research. *Environmental Education Research*, 8(3), 283–298.

Department of Conservation. (2016). *Marine Reserves A–Z*. [Online] http://www.doc.govt. nz/nature/habitats/marine/marine-reserves-a-z/ (Accessed 21 April 2017)

Eames, C., Barker, M., Wilson-Hill, F., and Law, B. (2009). *Investigating the Relationship Between Whole-School Approaches to Education for Sustainability and Student Learning* [Online] http://www.tlri.org.nz/sites/default/files/projects/9245_summaryreport.pdf (Accessed 21 April 2017)

Eames, C., Law, B., Barker, M., Iles, H., McKenzie, J., Williams, P., and Patterson, R. (2006). *Investigating Teachers' Pedagogical Approaches in Environmental Education that Promote Students' Action Competence*. [Online] http://www.tlri.org.nz/publications (Accessed 21 April 2017)

EMR. (2016). *Experiencing Marine Reserves*. [Online] http://www.emr.org.nz/(Accessed 21 April 2017)

Fleet, H. (1986). *The Concise Natural History of New Zealand*. Auckland, New Zealand: Heinemann.

Fletcher, S., Potts., J. S., Heeps, C., and Pike, K. (2009). Public awareness of marine environmental issues in the UK. *Marine Policy*, 33, 370–375.

Forest and Bird. (2016). *Marine Reserves – FAQ* [Online] http://www.forestandbird.org.nz/what-we-do/campaigns/we-love-marine-reserves/marine-reserves-faq (Accessed 21 April 2017)

Forest and Bird. (2017) Critically endangered sea lion dead in trawl net. [Online] http://www.forestandbird.org.nz/what-we-do/publications/media-release/critically-endangered-sea-lion-dead-in-trawl-net (Accessed 21 April 2017)

Friends of Taputeranga Marine Reserve. (2015). *Sea Urchin – Kina* [Online] http://taputeranga.org.nz/the-marine-life/invertebrates/sea-urchins-kina/(Accessed 21 April 2017)

Hase, S., and Kenyon, C. (2013). *Self-Determined Learning: Heutagogy in Action*. London: Bloomsbury Academic.

Heimlich, J. E., and Ardoin, N. M. (2008). Understanding behavior to understand behavior change: A literature review. *Environmental Education Research*, 14(3), 215–237.

Hungerford, H., Peyton, R. B., and Wilke, R. J. (1980). Goals for curriculum development in environmental education. *The Journal of Environmental Education*, 11(3), 42–47.

Hungerford, H., and Volk, T. L. (1990). Changing learner behavior through environmental education. *The Journal of Environmental Education*, 21(3), 8–21.

Jensen, B. (2002). Knowledge, action and pro-environmental behaviour. *Environmental Education Research*, 8(3), 325–334.

Jensen, B. B., and Schnack, K. (1997). The action competence approach in environmental education. *Environmental Education Research*, 3(2), 163–179.

King, M. (2003). *The Penguin History of New Zealand*. Auckland, New Zealand: Penguin Books.

Kollmuss, A., and Agyeman, J. (2002). Mind the gap: Why do people act environmentally and what are the barriers to pro-environmental behaviour? *Environmental Education Research*, 83), 239–260.

Lindsay, L. (2015). Transformation of teacher practice using mobile technology with one-to-one classes: M-learning pedagogical approaches. *British Journal of Educational Technology*, 47(5), 883–892.

Lloyd-Jenkins, D. (2016). *Beach Life: A Celebration of Kiwi Beach Culture*. Auckland, New Zealand: Godwit.

Lu, S., and Liu, Y. (2015). Integrating augmented reality technology to enhance children's learning in marine education. *Environmental Education Research*, 21(4), 525–541.

McKenzie-Mohr, D., and Smith, W. (1999). *Fostering Sustainable Behaviour: An Introduction to Community-Based Social Marketing*. Gabriola Island, BC, Canada: New Society Publishers.

Merchant, G. (2012). Mobile practices in everyday life: Popular digital technologies and schooling revisited. *British Journal of Educational Technology*, 43(5), 770–782.

Ministry of Primary Industries. (2009a). *The Quota Management System* [Online] http://www.fish.govt.nz/en-nz/Commercial/Quota+Management+System/default.htm (Accessed 21 April 2017)

Ministry of Primary Industries. (2009b). *The State of New Zealand's Fisheries* [Online] http://www.fish.govt.nz/ennz/Publications/Annual±Reports/Annual±Report±2009/Welcome/The±State±of±New±Zealands±Fisheries.htm (Accessed 21 April 2017)

New Zealand Geographic. (1989). Goat Island revisited. [Online] https://www.nzgeo.com/stories/goat-island-revisited/(Accessed 21 April 2017)

New Zealand Herald. (2016). Raw sewage pours through elderly couple's property. [Online] http://www.nzherald.co.nz/nz/news/article.cfm?c_id=1&objectid=11749139 (Accessed 21 April 2017)

Pachler, N., Bachmair, B., and Cook, J. (2010). *Mobile Learning: Structures, Agency, Practices*. Boston, MA: Springer US. [Online] http://link.springer.com/10.1007/978-1-4419-0585-7 (Accessed 21 October 2017)

Paço, A., and Lavrador, T. (2017). Environmental knowledge and attitudes and behaviours towards energy consumption. *Journal of Environmental Management*, 197, 384–392.

Palmer, J. (1995). Environmental thinking in the early years: Understanding and misunderstanding of concepts related to waste management. *Environmental Education Research*, 1 (1), 35–45.

Parliamentary Commissioner for the Environment (PCE). (2004). *See Change: Learning and Education for Sustainability*. Wellington, New Zealand: PCE.

Parliamentary Commissioner for the Environment (PCE). (2015). *Preparing New Zealand for Rising Seas: Certainty and Uncertainty*. Wellington, New Zealand: PCE.

Parsons, D.(Ed.). (2013). *Innovations in Mobile Educational Technologies and Applications*. Hershey, PA: IGI Global.

Parsons, M., Nalau, J., and Fisher, K. (2017). Alternative perspectives in sustainability: Indigenous knowledge and methodologies. *Challenges in Sustainability*, 5(1), 7–14.

Plankis, B. J., and Marrero, M. E. (2010). Recent ocean literacy research in United States public schools: Results and implications. *International Electronic Journal of Environmental Education*, 1(1), 21–51.

Räthzel, N., and Uzzell, D. (2009). Transformative environmental education: A collective rehearsal for reality. *Environmental Education Research*, 15(3), 263–277.

Salmond, A. (2017). *Tears of Rangi: Experiments Across Worlds*. Auckland, New Zealand: Auckland University Press.

Science Media Centre. (2011). Rena oil spill: Experts on environmental impact, clean-up. [Online] http://www.sciencemediacentre.co.nz/2011/10/11/rena-oil-spill-experts-on-environmental-impact-clean-up/(Accessed 21 April 2017)

Sea Change. (2017). *Hauraki Gulf Marine Spatial Plan*. [Online] http://www.seachange.org.nz (Accessed 21 October 2017)

Somekh, B. (2007). *Pedagogy and Learning with ICT: Researching the Art of Innovation.* New York, NY: Routledge.

Stapleton, S. R. (2015). Environmental identity development through social interactions, action, and recognition. *Journal of Environmental Education*, 46(2), 94–113.

Sustainable Coastlines. (2017). *About Sustainable Coastlines.* [Online] http://sustainable coastlines.org/about/overview/(Accessed 21 October 2017)

Thapa, B., Graefe, A. R., and Meyer, L. A. (2005). Moderator and mediator effects of scuba diving specialization on marine-based environmental knowledge-behavior contingency. *The Journal of Environmental Education*, 37(1), 53–68.

Thomashow, M. (1995). *Ecological Identity: Becoming a Reflective Environmentalist.* Cambridge, MA: MIT Press.

Traxler, J. (2010). Will student devices deliver innovation, inclusion, and transformation? *Journal of the Research Center for Educational Technology (RCET)*, 6(1), 3–15.

UNCED. (1992). *Agenda 21* [Online] www.unesco.org/education/English/chapter/chapter. shtml (Accessed 21 April 2017)

UNCSD. (2012). *The Future We Want.* Rio de Janeiro, Brazil: United Nations Conference on Sustainable Development.

UNESCO. (1975). *Belgrade Charter.* UNESCO. [Online] http://unesdoc.unesco.org/ images/0001/000177/017772eb.pdf (Accessed 27 May 2018)

UNESCO. (1978). The Tbilisi Declaration. *Connect: The UNESCO–UNEP Environmental Education Newsletter* 3(1), n.p.

Uzzell, D. (1999). Education for environmental action in the community: New roles and relationships. *Cambridge Journal of Education*, 29(3), 397–413.

Waikato Regional Council. (2016). *Mangroves.* [Online] http://www.waikatoregion.govt.nz/ PageFiles/1279/Mangrove%20factsheet.pdf (Accessed 21 October 2016)

Wals, A. E. J. (1994). Action taking and environmental problem solving in environmental education. In B. B. Jensen and K. Schnack (Eds.), *Action and Action Competence as Key Concepts in Critical Pedagogy.* Copenhagen: Royal Danish School of Educational Studies.

Wals, A. E. J., and Dillon, J. (2013). Conventional and emerging learning theories. In R. B. Stevenson, M. Brody and A. E. J. Wals (Eds.), *International Handbook of Research on Environmental Education.* New York: Routledge. pp. 253–261.

Warner, A., Eames, C., and Irving, R. (2014). Using social media to reinforce environmental learning and action-taking for school students. *International Electronic Journal of Environmental Education*, 4(2), 83–96.

Zeppel, H. (2008). Education and conservation benefits of marine wildlife tours: Developing free-choice learning experiences. *Journal of Environmental Education*, 39(3), 3–18.

Zeppel, H., and Muloin, S. (2008). Conservation benefits of interpretation on Marine Wildlife Tours. *Human Dimensions of Wildlife*, 13(4), 280–294.

6 History and heritage

Re-examining seascapes aboard the *Charles W. Morgan* (America's last whaling ship): a return to sea after 90 years

Nathaniel Trumbull

For more than 90 years the *Charles W. Morgan*, the world's last, and largely original, wooden whaling ship, stood at dock in the shallow waters of the Mystic River in the New England state of Connecticut. In a surprise to those who had visited the *Morgan* over the decades and knew the ship only as a dockside exhibit, the decision was made in 2009 by the Mystic Seaport, the 'Museum of America and the Sea', to re-outfit and take the *Charles W. Morgan* to sea once again (see Figure 6.1). The decision came as the *Morgan* was undergoing dry dock work and the condition of her hull proved to be in better shape than had been expected. The ambitious plan to take the *Morgan* back to sea was quickly embraced by those interested in supporting the Mystic Seaport and New England's whaling heritage. During the course of two months in the summer of 2014 the *Morgan* would sail to several traditional whaling ports of New England with the mission to 'make history come alive for today's audience' on what would come to be called the '38th Voyage' of the *Morgan* (Mystic Seaport, 2017a, n.p.).

No one aboard on the 38th Voyage had been alive the last time the *Morgan* had sailed. Ships still hunt for whales today, but the process is highly automated and arguably an entirely different experience than that of 100 and more years ago aboard a *sailing* whaling ship. The *Morgan* had traversed the globe on multiple trips to the Pacific from its homeport of New Bedford, Massachusetts over a period of 80 years of active whaling. Maritime communities of the New England region had once been at the epicentre of a worldwide whaling trade, reaching from the North Pacific to the coast of Antarctica. The 38th Voyage would become a tangible reminder to New England residents, and a lens of interpretation for those involved in the voyage, of the nation's maritime heritage (whaling ships' voyages were routinely referred to by their sequential number, regardless of their duration, which could range from months to several years).

For participants, and for significant numbers of the residents of coastal New England at large, the 38th Voyage came to create a series of engagements with 'past' seascapes via forms of civic involvement, exhibits and public programmes. Those seascapes focused on a specific historical period and took place in a multitude of ways for different participants, including engaging with the *Morgan* from on-board the vessel, to sightings of it at sea, to visiting aboard the *Morgan* during her port

Figure 6.1 Charles W. Morgan at Massachusetts Maritime in July 2014. Photo by the author.

calls. As Gillis (2012), Bélanger (2014) and Steinberg and Peters (2015) argue, the role of the sea, seacoasts and life at sea is only beginning to receive sufficient attention from a public perspective. Similarly, within a discussion of nostalgia, Day and Lunn (2003) point to the important role of providing a 'direct linkage' with the past in developing heritage studies and maritime history. 'One of the best ways of understanding this is to listen to the voices of people who have lived by and from the sea', they argue, citing the author of an oral history about the lives of the Orkney Islanders (Towsey, 2002, cited in Day and Lunn, 2003, 1). Maritime heritage is a salient example of a field where multiple understandings of the past are produced through different ways of materialising that past (Laurier, 1998).

Within the context of the heritage of vessels and museums, Easthope provides a useful scheme in distinguishing among the role of stationary exhibits, 'museum' vessels, sail training vessels and commercial vessels in shaping our views of maritime heritage (Easthope, 2001). He argues that restored or replica sailing vessels in particular can provide an 'authentic' heritage experience and points to the examples of the *Cutty Sark*, the *Victory*, the *Mary Rose* (in the UK), the *Constitution/Ironsides* and the *Våsa*. As this chapter will argue, the 38th Voyage of the *Charles W. Morgan* would go a step further in creating a more direct experience in terms of shaping views of maritime history, practices, traditions and way of life during a period when whaling was the most actively practised in history.

Eighty-five Voyagers, as they came to be called, became active participants in the *Morgan*'s 38th Voyage. The eyes and ears of each of those Voyagers, selected from a wide variety of creative professions, provided the narratives through which the public could learn from the 38th Voyage about whaling and nineteenth century life at sea more generally. The Voyagers each employed different approaches and techniques and collectively created an assemblage of artistic reports on the Voyage based on their twenty-first century experiences aboard a nineteenth century sailing vessel. Indeed, the website on which the Voyagers' creative work was displayed for each of the legs of the *Morgan's* journey resembles a quilt.[1] Other participants in the *Morgan*'s 38th Voyage would create seascapes through shoreside opportunities for interaction with the 38th Voyage or vicariously through press reports of the Voyagers' experiences.

Interest in the modern-day voyage of the *Morgan* was high, especially in those communities where the *Morgan* would make port calls during her two-month voyage. Never before had modern-day residents in these coastal communities of New England been able to see so directly and tangibly – on their own water-fronts – such a large and visible living artefact of nineteenth and early twentieth century whaling. Residents could climb aboard the *Morgan* at dockside and witness first-hand some of the tools and practices of a whaling ship that had travelled under her own sails and into their bays and harbours. In this way, residents of New London, Newport, Martha's Vineyard, New Bedford, Province-town, Boston and Massachusetts Maritime experienced the *Morgan* directly. Several of those town and cities can trace their meteoric economic rise in nineteenth century America directly to the enormous profits of the whaling industry (Dolin, 2007).

In this chapter, I explore the experiences of the 38th Voyage of the *Charles W. Morgan* as they illuminate maritime heritage not only from the perspective of the Voyagers who took direct part, but also from the perspective of the many other observers and indirect participants who learned about whaling, whaling ships and life on-board from the 38th Voyage. This chapter unfolds with an overview of the setting of the whaling industry in New England, a brief account of how the 38th Voyage was conceptualised, before delving into the direct experiences of the Voyagers aboard the 38th Voyage and exploring the public perceptions of the Voyage from the perspective of maritime heritage and history.

Contextualising the *Morgan*

The rapid growth of the whaling industry and its impact on the way of life were hallmarks of newfound wealth for a significant number of New England coastal communities in the late eighteenth and nineteenth centuries. The industry brought untold riches to a region that had previously been a relatively minor economic actor on the world stage. Maritime communities on the islands of Martha's Vineyard and Nantucket, as well as in the mainland cities of New Bedford, New London and Salem perhaps benefited most directly from the newfound wealth of whaling. Their historic downtowns today contain architectural 'gems' from that period of new wealth. They became among the wealthiest communities of America at the time, rivalling if not surpassing some of the longer-established wealthy neighbourhoods of Boston, New York City and Philadelphia. At whaling's peak, 2,700 whaling ships plied the world's oceans in search of whale oil and baleen (used for, among other things, women's clothing). Whales were known during the period as the primary source of illumination and lubrication. The latter role has not been entirely lost; space missions continue to make use of small amounts of whale oil for this purpose. Petroleum products would later become the modern equivalent for the functioning of everyday life in many ways (Yergin, 1991).

The hard labour of whaling, the multi-year absences of whaling ships' crews, and the always-present danger that a ship might not return, became not only part of the economic ascendance of those regions, but also a key cultural identification that is celebrated to this day. Many residents of the New England region claim or can trace familial, or at least cultural, roots to the sea. My own parents gave me my second name, Starbuck, directly out of the American maritime novel, *Moby Dick.* The novel's crew sails out of Nantucket, and while neither of my parents ever lived there, the location held some cultural significance for them as the place they were married.[2]

There are other parallels between whaling and modern-day boating in these maritime communities. One might argue that in a similar way to how the whaling trade spun together networks of investment, capital and profits, today's owners of sailboats in the region mix (often globally sourced) commerce with their leisure activities. Indeed, some of the world's most expensive sailing ships ply the waters off Newport, Martha's Vineyard and Nantucket for purposes of pleasure and sport. Owners of yachts in those communities today can be fabulously wealthy and maintain a dozen crew members full-time throughout the year for the sole purpose of participation in a handful of sailboat races. In the case of the islands of Nantucket and Martha's Vineyard, one might even argue that the maritime global economic activity of whaling was a distant precursor to modern-day recreational sailing off those islands. Funded by profits earned in distant locations of the nation or globe, those large pleasure yachts are the most obvious identifying landmarks today of wealth on the waterfront, just as whaling ships and their captains' family residences once were (Dolin, 2007).

The *Charles Morgan*'s 38th Voyage was bound to attract special attention, no matter how largely unexceptional the *Morgan*'s past had been. The *Morgan*'s origins were common. She had been built in New Bedford as a standard ship of the period and designed for whaling. She was constructed in less than a year. Timber for the ship came from New England, as did its builders. The *Morgan* was launched in 1841. Its crew would evolve significantly during her next 80 years of service. Crews in the beginning decades were locally based young men, whose ancestors and relatives would also have been involved in the whaling trade. During the ship's last decades, migrant crews would come to handle the ship. Bolster and others have written in detail about the socio-economic plight of some of those later-period whaling ship crews, especially toward the end of the heyday of profits and dominance of the trade by New England based ships

(Bolster, 1997). The difficult nature and lives of those crews provides a stark contrast to the glorified stories of whaling and wealth of the period. While we may today associate the sealing and whaling era with such early celebrated trips as those of Captain Nathaniel Palmer of Stonington, Connecticut, grave dangers to the crew, both financial and mortal, were never far from the surface in the whaling world (Palmer, 1986). More than one whaling ship captain's fate led to bankruptcy when whales were few, or selected routes proved to be unlucky. Ships could founder at a location on the other side of the globe. Labour disputes among crew and the captain could erupt into violence that might disrupt and jeopardise months and years of travel across the world. The increasingly intensive whaling activity on a global scale would lead, only a few decades later, to near-extinction levels of some whale populations. It was against this backdrop of local, regional and global maritime heritage, including at least some nostalgia, but also of profound violence of the epoch of whaling industry, that the Mystic Seaport contemplated sending the *Morgan* back to sea.

The Mystic Seaport's plan

In designing the 38th Voyage the Mystic Seaport aimed to address a wide range of reflective questions from the perspective of a modern-day maritime observer. Museum staff, Voyagers and the public at large were of course very curious about the mechanics of how a nineteenth century whaling ship would sail with a twenty-first century crew. The vessel was essentially a time machine in this regard. But no one doubted that a direct experience of sailing the *Morgan* would also yield new insights into life aboard a whaling ship. Many related questions arose. Could modern-day artists, scientists and other scholars at sea as Voyagers learn – each from the perspective of their own disciplines – about the maritime experience of a century and half ago by each spending time aboard a different leg of the *Morgan* during her 38th Voyage? Might those Voyagers also learn something new about ourselves as modern-day observers? As Nicol has written about the value of direct experiences on the sea, 'for humans to better understand their interconnection with nature, they need to have direct encounters' (Nicol, 2015, 246).

While the Mystic Seaport leadership's decision to take the museum's most valuable artefact into its natural element for the first time in many decades represented a carefully calculated risk, the potential benefits of bringing that artefact to the public over a 500-mile distance of a relatively densely populated coastline were a powerful incentive to pursue the Voyage. While the *Morgan* had been turned bow to stern regularly over the years at the dock, she had not moved under her own power for 90 years. Hundreds of thousands of visitors have set foot on the *Morgan* over the past decades as visitors to the Mystic Seaport Museum, but the 38th Voyage was envisioned as an entirely different experience. A total of 85 artists, historians, scientists, journalist, teachers, musicians, scholars, including a handful of whaling family descendants, would eventually take turns on different legs of the trip aboard the *Morgan* on its two-month 38th Voyage.

From an institutional perspective, the Mystic Seaport has the goal of designing creative products to share with the public. While participation in the 38th Voyage was to be free of charge, a selection committee sought to identify individuals whose backgrounds could help them to serve as narrators of the 38th Voyage experience in a variety of different media forms. The 85 individuals who would become the Voyagers would be asked to serve not only as ambassadors of the ship, but also to use their creative skills, perspectives and talents to share their experience both within their own professional circles, but also with the public at large. 'While rooted in history, the 38th Voyage was not a reenactment, but an opportunity to add to the *Morgan*'s story with contemporary perspectives', explained the Mystic Seaport (2014). By selecting a wide range of Voyagers of differing perspectives, the Mystic Seaport sought to examine broadly the past experiences of those who had once sailed the ship, but also to relate those experiences to a contemporary audience.

An academic-oriented charrette organised two years before the start of the 38th Voyage had brought together some of the most recognised scholars in the field of American maritime history, maritime literature, maritime archaeology, marine and whale science (including Helen Rozwadowski, Jeffrey Bolster, D. Graham Burnett, Lisa Norling, Joe Roman, Nancy Shoemaker, Tim Runyan and Elizabeth Schultz). Recommendations from these scholars were incorporated into many of the preparations for the Voyage, including the *Morgan*'s dockside exhibits. The topics of those exhibits were to include the essential role of African-Americans and native Americans in whaling, other social and cultural aspects of the whaling trade, relevant scientific topics related to whaling, and even the condition of whales in our oceans today. Partnerships with external organisations and agencies was extensive both in preparation of the exhibits and during the Voyage itself. For example, the New Bedford Whaling Museum, USS Constitution Museum, National Oceanic and Atmospheric Administration/Stellwagen National Marine Sanctuary staff and Massachusetts Maritime Academy educators were all key partners of the Voyage (NOAA, 2014a).

The time frame for each of the Voyagers was to sail aboard for one leg, generally defined as one night in the ship's berths and then a single day sailing (for purposes of safety, the *Morgan* did not sail at night during the 38th Voyage; this schedule broke entirely with what would have been the standard sailing route of the first 37 voyages). Some of those legs were more mundane than others; a leg from Martha's Vineyard to New Bedford, for example, involved a minimal distance (a few dozen kilometres) in the largely protected ocean in the lee of the Elizabeth Islands. Other legs would prove to be more iconic, including sailing among humpback whales off Provincetown (an experience that tourists on whale watching trips can replicate today, though of course from a quite different deck). Due to the ship's status as a National Historic Landmark, the *Morgan* spent each night safely berthed at dockside. As weather conditions were a determining factor in the decision to head to sea each day, each port transit was scheduled with a three-day window of departure, with the goal being that the ship would sail on the first acceptable day. Once safely in port, the *Morgan* was open to the public at select times and ordinarily every day. Additionally, the ship was accompanied by an extensive dockside exhibition that included exhibits of historic interpretation, live demonstrations and other related waterfront activities (a large inflatable whale was one of the unmistakable hits for young visitors).

Collectively the Voyagers produced dozens of different means and formats of interpretations of the 38th Voyage. Each Voyager worked in their own medium, whether it be through writing, visual arts or other forms. American writer Herman Melville had intentionally joined the crew of a nineteenth century whaling ship in order to learn more about a crew member's experience aboard a whaling ship and then to write about it. His explicit goal had been to build up a series of experiences on which he could then draw in his writing, which became most famously the basis for *Moby Dick*. Similarly, Voyagers were aboard to collect impressions through experiences and then, by sharing those experiences in their own preferred medium, become ambassadors of the 38th Voyage to the public at large.

Preparing the vessel

The *Morgan*'s fate had hung in the balance many times, especially after extensive damage to her hull in the early twentieth century while berthed in the port of New Bedford. A series of entrepreneurs had then recognised the potential value of the ship as a destination for tourists as the single last example of a wooden whaling ship. But resources to take care of the *Morgan* did not materialise until the Mystic Seaport purchased her in 1941. Repair to the ship was expensive and problematic. Indeed, one of the first investments in what would become the 38th Voyage was a multi-million dollar upgrade to the Seaport's mechanical lift for bringing the *Morgan* out of the water for extensive repairs.

A five-and-a-half-year, 7.5 million dollar restoration of the ship ensued, during which the *Morgan* became a sailable vessel. One of the discoveries of the repair was the remarkable condition of the original keel. The shipwrights of the Mystic Seaport, of whom a small team is on the permanent staff, were themselves surprised at how intact the keel turned out to be. When new boards were needed, 'authentic' sources for that timber were found. Many of the new planks of the *Morgan* would come from southern live oak salvaged from Hurricanes Katrina and Hugo. Another source of the timber for the ship repairs came from a construction company in the Charlestown neighbourhood of Boston. In clearing a pond 80 feet below sea level to prepare a new hospital foundation, workers had found ship lumber originally placed in the pond for future shipbuilding. The timber even had its original numbers. Eighteen truckloads of that timber were donated to the restoration of the *Morgan*. Nevertheless, it is estimated that only about 15 per cent of the boards on the ship are original. Whether this number is low or high depends on the eye of the beholder. For a wooden shipbuilder, such a small percentage may seem remarkably high. The intertwined old and new materialities might well be understood as a symbol of the ship as an instrument of engagement in the present with the past.

The Voyage through the eyes of artists, scholars and others, and their creative products

Voyagers 'overnighted' in the original berths of those of nineteenth century sailors. In many ways they were much freer than nineteenth century crew members would have been. They were generally able to go where they chose on the ship. They were also free from the actual work of the running of the ship. The Mystic Seaport established certain rules for the Voyagers. They did not handle the lines or sails, but were left to observe and engage in their own activities, artistic and otherwise. That said, the Voyagers slept in the forecastle of the ship like common hands and each took a turn at the helm. A dozen Mystic Seaport crew members were fully responsible for the sailing of the ship, a full half of whom were female (in a clear shift from nineteenth century tradition). This shift in gender of crew did not go unnoticed by the Voyagers. As Blum observed,

> [m]y admiration and covetousness in watching the crewwomen of the *Morgan* work was self-identification, in part; once I too was young and strong and free of orthopedic injury … The presence of these women on ship, I realize now, had a different effect on me than an all-male crew.
>
> (Blum, 2014)

In observing and engaging in their own activities some Voyagers sketched, others took notes, and others took photo, video and audio recordings.[3]

Indeed, after their on-board experience, each of the Voyagers submitted one or more creative products from the Voyage. Each Voyager in this way brought their own lens to the Voyage – that is, their own professional perspectives, through which they perceived the sights and sounds of activities on the Morgan and at sea. For example, one Voyager wrote original music that blended the folk traditions and sea shanties that might have been heard aboard the *Charles W. Morgan* in the nineteenth century (Wikfors, 2014). Dante Francomano composed a musical piece titled *Cetacean Citations*, 'a contemporary concert piece inspired by the soundscapes of whales and whaling and by sounds recorded during the Morgan's 38th Voyage' (Francomano, 2014). Another voyager wrote a graphic travel log, or in her own words, 'A travelogue comic documenting my trip to Boston, my time sailing aboard the vessel, and my reflections on tall ship culture, authenticity, and adventure' (Bellwood, 2014). Similarly, a painter aboard explained: '[o]ne overnight and a full day on an authentic whaling ship, on the sea surrounded by whales … the intangible things of which paintings are made' (Hanson, 2015). Some Voyagers focused on simply observing the crew at work. As Blum writes, '[b]earing witness to the crew's labors did more for my nautical knowledge than would four years with a carefully curated shipboard library (or four times that with a university library)' (Blum, 2015).

The inclusion of a 'Stowaway' Voyager, chosen to represent the 'everyday person' was also a feature of the trip. The 'Stowaway' was chosen from a large pool of candidates after a well-publicised national search and attracted wide attention in the press. Ryan Leighton, a journalist from Boothbay, Maine, served as the 'Stowaway' voyager, and unlike the other Voyagers, was aboard for all legs of the 38th Voyage. As the Stowaway, Leighton transmitted video and posted actively about the Voyage on social media, reaching an audience that came to encompass not only those communities through which the *Morgan* passed, but all those to whom his impressions were posted and then reposted. In this way he became one of the more visible observers of the entire voyage, as he delivered news and impressions of the Voyage almost daily. He was often aloft in the rigging with his camera. The Stowaway perhaps had the most license, if not mandate, to inject twenty-first century views and insights into the Voyage. His blogging was often humorous. He described how he had had to move to sleep in a hammock in the very high humidity of the hold, in large part to distance himself from snoring in the 'fo'c'sle'. He described interesting and unusual places that he visited during his time onshore at the different ports. He even included a celebrity sighting, or more precisely participation, when David Letterman dropped by the *Morgan* in Vineyard Haven on Martha's Vineyard and took one of the whaling boats for a row. Leighton concludes at the end of the trip, 'the 38th Voyage of the *Charles W. Morgan* has felt like sprinting a marathon. It has been two months of exhilarating, inspiring, exhausting, important, sleepless, sweaty and spectacular moments, with a free soda thrown in from time to time' (Leighton, 2014).

The 38th Voyage's trip images and reflections from Voyagers on their experiences aboard are now online.[4] The Mystic Seaport created a large multimedia online repository for the 38th Voyage that allows viewers to explore the stories, images, scientific data, artistic impressions and artefacts collected by the Morgan Voyagers. As a senior member of the Mystic Seaport commented after the 38th Voyage's completion, 'we now have a powerful body of knowledge, sensory experiences, images, sounds, and visceral and artistic human responses that all contribute to our understanding of nineteenth century whaling and the human-whale dynamic' (Funk, 2014).

Impressions from dockside

A total of approximately 64,000 visitors experienced directly some element of the 38th Voyage from onshore (Figure 6.2). The dockside visits became a fundamental part of the overall Voyage in the sense that interpretation and thinking about each of the legs of the trip continued there. According to Rocky Hadler, third mate, 'One of the things I enjoyed the most is when you'd see people step on the deck, whether they'd ask any good questions or not, but all of a sudden you'd see this smile' cause [they] could feel like she was alive. That was cool' (Mystic Seaport, 2015, n.p.).

Figure 6.2 Charles W. Morgan at Massachusetts Maritime in July 2014. Photo by the author.

I was present at two of the dockside visits, in Vineyard Haven on the island of Martha's Vineyard and at the Massachusetts Maritime Academy (Wareham, Massachusetts), as well as the final return to Mystic. Just as the sight of the *Morgan* in the centre of each of the towns or cities was visually captivating, so were the dockside visits remarkably engaging. As Karen Wizevich wrote in the *Summative Evaluation of the Charles W. Morgan Dockside Experience* – a survey she conducted of the dockside experience of the *Morgan* in New London, New Bedford and Boston – she pointed to the 'clear symbiotic benefits between the *Morgan* and the rich interpretive context represented by the dockside experience' (Wizevich, 2014, 1). The visit of the *Morgan* to each of its ports of call was likely among the foremost cultural events of the summer for those communities. In addition to the dockside exhibit visitor experience, visitors in each of the locations could attend public lectures, films and other related educational events.

The official number of visitors to the dockside visits of the *Morgan* of course does not capture the much larger numbers who observed the Voyage on the web through video and news stories, on television or in the local press. The extraordinarily rich online offerings (as a geographer, I was particularly taken by the real-time automatic identification system (AIS) mapping of the *Morgan's* trip on the Voyage's website) were highly engaging. The large number of residents and visitors who went to nearby beaches or coast to view the *Morgan* as she passed by under sail was also remarkable.

My own personal 'sighting' of the *Morgan* under full sail, in the waters off my childhood home in Woods Hole, Massachusetts, looking toward West Chop (located within the Town of Aquinnah, recently renamed after the local native American tribe), was truly unforgettable. The ship occupied so much more of the horizon, stood so much higher, than any other ship I had ever seen on the horizon of that body of water. One could only imagine what a nineteenth century horizon in the same location might have looked like with multiple such ships on that body of water. This view was all the more astonishing as this is a horizon with which I have been familiar since childhood. My reflections on viewing the *Morgan* under sail off this part of the New England coast are of course part of a quite local and personal situated knowledge, involving many other memories of this location.

Environmental interests and ironies: new perspectives on whaling

One of the more publicised legs of the 38th Voyage in the media took place off the town of Provincetown, located on the most northern point of Cape Cod, in the Stellwagen Bank National Marine Sanctuary. While the site is often the destination of whale watching boats and tourists from both Cape Cod and Boston (fast whale watching boats can reach the Sanctuary in about an hour from Boston), that day the *Morgan* sailed off Provincetown included an extensive encounter with several local humpback whales. While the scene could not have been exactly scripted in advance, its chances were indeed enhanced by the choice to include the Stellwagen Bank on the trip's itinerary. Two small whaleboats from the *Morgan* would come remarkably close to several of the humpback whales. While no harpoons were thrown – or were even on board those whale-boats – the oarsmen of those small boats perhaps had a better sense of how the so-called Nantucket sleigh ride might begin. That is, they could likely imagine the moment when a recently harpooned whale dives, in an effort to throw off the harpoon, and the whaleboat is towed forwards violently. Sean Edwards, who had been aboard one of the whaleboats, in fact expressed his concern to me later that the *Morgan* whaleboats had been so close to the humpback whales that the *Morgan* might have been be in violation of the law against interfering with a federally listed endangered species (NOAA, 2014a).

Natural scientists were also involved in an early stage of the organisation of the Voyage and were careful to portray the *Morgan*'s mission within the context of modern-day environmental sensibilities and concerns. As part of NOAA's OceanLIVE online broadcast, oceanographer Sylvia Earle joined the Voyage on one of the days in the sanctuary and called the *Morgan* a 'ship of hope' in that it could represent the cause of protection of our oceans and its creatures. The OceansLIVE (real-time) broadcast offered interviews and commentary with historians, scientists, authors and artists discussing the shift from whaling to watching in New England during some legs of the Voyage. Jean-Michel Cous-teau joined Earle for one of those live broadcasts from Provincetown (NOAA, 2014b). Other scientists sought to engage in scientific work directly from aboard the *Morgan*. Michael Whitney of the University of Connecticut made measure-ments of weather aboard and constructed a web tool linking weather conditions of the current day to nineteenth century weather reports. He also launched a number of 'drifters' with GPS units in order to track ocean surface currents as they passed by the *Morgan*.

The Mystic Seaport and the Voyage had been understandably concerned that celebrating a vessel whose sole function had been to hunt whales might propagate a poor message of environmental awareness and conservation to a twenty-first century public. On the other hand, one could argue that the *Morgan* provided the most relevant possible opportunity to explore the past (and current situation in a few remaining parts of the world) of the violence of whale hunting. As the Seaport explained in a press release about sending the *Morgan* to the waters of the Stellwagen Bank National Marine Sanctuary off Provincetown of Cape Cod,

> this time the goal was to raise awareness of the fragile state of our oceans and how important they are in our ecosystem. Sailing an artifact of a defunct – yet once very important – industry among the creatures it sought to kill, offers an opportunity to compare current practices and technology with the past. Both the *Morgan* and the whales have survived and there are lessons in that survival.
>
> (Mystic Seaport, 2017b)

Another way to view the legacy of the *Morgan*'s 38th Voyage was that it had been brought to sail among whales once again, but this time peacefully.

Port calls and homecoming

A visiting ship's port calls are filled with much expectation and meaning both for crew and those with whom the visit takes pace. As a child at school in Woods Hole, Massachusetts – today the site of the Marine Biological Laboratory and Woods Hole Oceanographic Institution – I can remember the coming and going of Soviet oceanographic vessels (remarkable then, and even more remarkable to consider today from the perspective of current US–Russian relations). In the 1970s, it was perhaps my first exposure to other ways of life at the end of the horizon. The appearance of Soviet research vessels and of their scientists on the village streets of Woods Hole is a lasting childhood memory. They brought Soviet candy for the schoolchildren and showed their own cartoon movies in our elementary classroom (a medium which quite effectively broke the language barrier). The scientists and crew had been at sea for long periods, usually months, and were eager to engage with residents of the Woods Hole community. Only in later years would I hear the accounts of how Soviet and Polish fishing fleets, among others, had played a role in overfishing the rich Georges Bank fishing grounds off New England. This led in part to the creation of the 200-nautical mile Exclusive Economic Zone and a subsequent ban on fishing by foreign fleets off the New England coast.

The *Morgan* thus shared a part of itself, its identity, with those New England cities and towns it visited in the summer of 2014. One of the more symbolic port of calls came in Boston where the *Morgan* docked next to the

USS Constitution, affectionately known as 'Old Ironsides' and the world's oldest commissioned naval vessel. It would have been difficult for any visitor to the *Constitution* – which daily welcomes hundreds of visitors to the Charlestown Navy Yard and is today part of the Boston National Historical Park in Boston – not to have seen the symbolism of a young revolutionary nation's naval presence in the *USS Constitution* and the accompanying *Morgan*, a later reflection of a growing nation's global economic reach in the realm of the sea.

The *Morgan* completed her historic 38th Voyage on 6 August 2014, a little over two months after her return to open sea. I was in attendance at the unofficial homecoming ceremony of the *Morgan* back to Mystic Seaport upon the end of her trip. I had never seen the Mystic River so full of small boats, shoreline spectators and well-wishers, with its raised drawbridge as the *Morgan* returned home. During the impromptu homecoming (or relatively impromptu, as the *Morgan*'s 38th Voyage had been scripted right down to the smallest detail, including the final words of its crew and captain on the dock of the Mystic Seaport), the words of one of the speakers rang especially true: 'It will likely be years for the fuller interpretation and lessons of the 38th Voyage to be realised'.

Indeed, so much material was gathered, not only in terms of creative products of the Voyagers, but also from a technical perspective, that researchers will be revisiting and analysing the experience of the Voyage for many years to come. One of the more commonly overheard comments from the crew was how well the ship had sailed, to their surprise and delight. She sailed much further into the wind than they had anticipated. She was also much faster than they had expected her to be (this speed indeed led to more than one early arrival in port on the different legs of the 38th Voyage, sometimes to the confusion of observers who had been given a later time to come and observe the vessel pass by or arrive along the coast.[5]

The knowledge gained during the sailing component of the 38th Voyage will continue to be available, especially through the stories and narrative of the Voyagers. The inclusion of the Voyagers and their mission to provide their own interpretations of the Voyage was of course a stroke of genius on the part of the Mystic Seaport. Involving the Voyagers with life aboard the *Morgan*, including the obligatory night on the ship before their leg of the voyage, changed everything in terms of how the story of the *Morgan*'s 38th Voyage would come to be told. As a crew member (Rocky Hadler, third mate) explained, 'If you're walking onto a ship that's moving, that's sailing, you look at it with different eyes' (Mystic Seaport, 2015). Today the *Charles W. Morgan* is docked back at Chubb's Wharf at Mystic Seaport and has resumed her role as an exhibit and the flagship of Mystic Seaport. There is talk of future voyages for the *Morgan*, but given the scope of the effort involved and financial constraints, it seems unlikely to happen again in the immediate or even near future. The Mystic Seaport's latest initiative has been more land-based –the opening of its now largest and newest exhibition space, the Thompson Exhibition Building

Conclusions

The statement '[r]eturning the last remaining wooden whaleship to sea will be one of the most evocative experiences of our generation' had sounded grandiose before the 38th Voyage, but there was an element of truth to it (Mystic Seaport, 2013). In an extraordinary feat, a nearly fully functioning nineteenth century whaling ship had been observed by those on shore and witnessed firsthand from the perspective of modern-day artists, scientists and other scholars. The 38th Voyage permitted a twenty-first century mindset to directly engage with experiences aboard a nineteenth century ship. Through a series of engagements with its heritage, the 38th Voyage unlocked past and contemporary reflections of how we once lived, currently live, and might possibly live moving forward, with the sea.

It seems, for example, that the 38th Voyage did play a role in changing contemporary New England residents' views of nineteenth century whaling ships, of life aboard those ships, and of their ancestors' maritime communities. The Voyage was a front-page story in the newspapers of those cities and towns the *Morgan* visited. Certainly, the experience of the 38th Voyage was a memorable one for coastal residents of the region. The Voyage also represented the opportunity for new representations and portrayals of a series of seascapes of American maritime history not only for 38th Voyage participants, but for tens of thousands of other observers of the Voyage who encountered it from the shore, and from further afield through globalised media technologies.

By the end of the 38th Voyage, the twenty-first century witnesses had been given the opportunity to reinterpret nineteenth century whaling experiences from a broadly different set of perspectives than before. Moreover, for Voyagers themselves the experience was profound. Those aboard the *Morgan* will likely continue to struggle with the consistency and mutability of their own narratives based on their single voyage on the whaling ship. One cannot doubt that a return trip, under different conditions of weather or on a different ship, with a different crew or other Voyagers would result in another series of interpretations, some perhaps competing with one's original interpretations. As one voyager wrote,

> [a]s I learned quickly in my short time on the *Morgan*, even the most skilled nautical professionals were constantly massaging the narratives of their expertise, of the ship's authenticity ... For Ishmael, as well as for the 38th Voyagers, whaling and poetry commingle. I don't yet know in what register my own *Morgan* memories will settle.
>
> (Blum, 2014)

By focusing on the embodied experiences of the Voyagers with the sea, the 38th Voyage resonated strongly for all those involved; whether Voyager, shoreside participant or vicarious observer. Re-enactments such as that of the 38th Voyage are vital for us to unpack and examine. The 38th Voyage was both carefully orchestrated and spontaneously curated by the Voyagers. That Voyage was indeed a more 'direct linkage' (Day and Lunn, 2003) to the past than a static museum piece could likely ever provide. Within Easthope's classification and schema of possible instruments for exploring maritime heritage, direct experiences such those of the Voyagers must surely be rated highly from the perspective of authenticity.

Maritime heritage helps to shape contemporary understandings of the sea by providing a lens through which today's observers can experience first-hand a critical period of the human experience with the sea. What was special about the *Morgan*, unlike other projects, was that it set sail and that it engaged participants across different spaces, scales and times – both directly and from a distance. The 38th Voyage of the *Charles W. Morgan* taught us the importance of making direct connections with the sea in order to understand our past, present, and envisage a better future for our relationship with the sea.

Notes

1 See the website: https://www.mysticseaport.org/voyage/nl-to-npt/
2 Howard Schultz also found in *Moby Dick* his future coffee company's name – the more famous 'Starbucks' (though he added an 's' to the end of the first mate's name in the novel.)
3 The National Endowment for the Humanities became one of the principal sponsors of the Voyagers.
4 See https://www.mysticseaport.org/voyage/voyagers/
5 I too encountered this problem of the *Morgan* running ahead of schedule when she passed between the Elizabeth Islands from Martha's Vineyard to New Bedford, well in advance of the scheduled time, and I nearly missed viewing her passage under full sail.

References

Bélanger, P. (2014). The other 71 percent. *Harvard Design Magazine*, No. 39, Wet Matter. [Online] http://www.harvarddesignmagazine.org/issues/39/the-other-71-percent (Accessed 18 March 2018).

Bellwood, L. (2014). *Down to the Seas Again*. [Online] https://lucybellwood.com/down-to-the-seas-again/ (Accessed 18 March 2018).

Blum, H. (2014). 18 hours before the mast. In: *Los Angeles Review of Books*, August. 3 https://v2.lareviewofbooks.org/contributor/hester-blum/#! (Accessed 18 March 2018).

Blum, H. (2015). 38th Voyage. [Online] https://sites.psu.edu/hester/38th-voyage/ (Accessed 18 March 2018).

Bolster, W. J. (1997). *Black Jacks: African American Seamen in the Age of Sail*. Cambridge, MA: Harvard University Press.

Day, A., and Lunn, K. (2003). British maritime heritage: Carried along by the currents? *International Journal of Heritage Studies*, 9(4), 289–305.

Dolin, E. (2007). *Leviathan: The History of Whaling in America*. New York: W. W. Norton.

Easthope, G. (2001). Heritage sailing in Australia: A preliminary schema. *International Journal of Heritage Studies*, 7(2), 185–190.

Francomano, D. (2014). Cetacean citations. [Online] https://www.mysticseaport.org/voyage/boston-to-mma/dante-francomano-cetacean-citations/ (Accessed 18 March 2018).

Funk, S. (2014). Whaleship celebrates whales in Stellwagen Bank Sanctuary. *Stellwagen Bank E-Notes*, July–August. [Online] https://nmsstellwagen.blob.core.windows.

net/stellwagen-prod/media/archive/library/pdfs/enotes_august2014.pdf (Accessed 18 March 2018).

Gillis, J. (2012). *The Human Shore: Seacoasts in History*. Chicago: University of Chicago Press.

Hanson, W. G. (2015). Sailing aboard the Charles W. Morgan. *ASMA News and Journal*, Winter 30. [Online] https://americansocietyofmarineartists.wildapricot.org/resources/Images/2015%20WINTER%20Lo-Rez.pdf (Accessed 18 March 2018).

Laurier, E. (1998). Replication and restoration: Ways of making maritime heritage. *Journal of Material Culture*, 3(1), 21–50.

Leighton, R. (2014). *Stowaway: P-Town to Beantown*. Online.

Mystic Seaport. (2013). The 38th Voyage – Spirit of American enterprise. [Online] http://38thvoyage.org/generation (Accessed 18 March 2018).

Mystic Seaport. (2014). The 38th Voyage of the Charles Morgan. [Online] https://www.mysticseaport.org/voyage/ (Accessed 18 March 2018).

Mystic Seaport. (2015). Crew interviews. [Online] https://www.mysticseaport.org/voyage/mystic-seaport/crew-interviews/ (Accessed 18 March 2018).

Mystic Seaport. (2017a). The 38th Voyage. [Online] https://www.mysticseaport.org/explore/morgan/38th-voyage/ (Accessed 18 March 2018).

Mystic Seaport. (2017b). Places. [Online] www.mysticseaport.org/voyage/places/ (Accessed 18 March 2018).

National Ocean and Atmospheric Administration, National Marine Sanctuaries. (2014a). *Charles W. Morgan Sails, Stellwagen Banks E-Notes*. June.

National Ocean and Atmospheric Administration, National Marine Sanctuaries. (2014b). The 38th Voyage of the Charles W. Morgan. [Online] http://sanctuaries.noaa.gov/whales (Accessed 18 March 2018).

Nicol, R. (2015). In the name of the whale. In M. Brown and B. Humberstone. (Eds.), *Seascapes: Shaped by the Sea. Embodied Narratives and Fluid Geographies*. Farnham: Ashgate, pp. 141–154.

Palmer, E. (1986). Young Cap'n Nat of Stonington. *The Mariner's Mirror*, 72(1), 25–42.

Steinberg, P., and Peters, K. (2015). Wet ontologies, fluid spaces: Giving depth to volume through oceanic thinking. *Environment and Planning D: Society and Space*, 33, 247–264.

Towsey, K. (2002). Orkney and the sea: An oral history, Orkney: Orkney Heritage. U.S. Senate Special Resolution 183 — Commemorating the Relaunching of the 172-year-old Charles W. Morgan by Mystic Seaport: The Museum of America and the Sea on July 17, 2013. [Online] https://www.congress.gov/bill/113th-congress/senate-resolution/183 (Accessed 20 March 2018).

Wikfors, G. (2014). Music for Waldzither. [Online] https://www.mysticseaport.org/voyage/stellwagen/stellwagen-11/gary-wikfors-music/ (Accessed 18 March 2018).

Wizevich, K. (2014). *Summative Evaluation of the Charles W. Morgan* Dockside Experience.: Mystic Seaport, unpublished report.

Yergin, D. (1991). *The Prize: The Epic Quest for Oil, Money, and Power*. New York: Simon and Schuster.

7 Sensory autoethnography

Surfing approaches for understanding and communicating 'seaspacetimes'

lisahunter

Known for his aquatic attentions, Steinberg (2014) has argued a challenge in making sense of and communicating our experiences of living with the seas. He reinforces that sea-writing, or thalassography, is difficult '[s]ince the sea is a space that cannot be located and that cannot be purely experienced' (Steinberg, 2014, xv). The sea is forever represented (in maps, books, films, paintings, poetry) but is also, ultimately *beyond* representation. To engage directly with the sea is an experience that cannot be neatly represented. This creates a quandary in how to understand that engagement, and communicate it. But why should we even care to understand and communicate? This chapter argues that such efforts are vital in order to guide our ethical sensibilities and responsibilities to the sea.

To begin to address the challenge or quandary of making sense of our watery worlds, literature about affect, embodiment and the senses (Anderson, 2009; Barad, 2003; lisahunter and emerald, 2016a; Pink, 2013; Sparkes and Smith, 2012) and human relationships with the sea (Amadio, 1993; Anderson and Peters, 2014; Brown and Humberstone, 2015; Peters, 2015; Steinberg, 2001) are vital. This chapter illustrates a modest attempt at what might loosely be called a sensory (auto)ethnography that may assist in attunement to thalassography. In what follows, I use a variety of practices to create *field texts*, to document an embodied memory and affect of an aquatic assemblage through the personal example of surfing.

As a means to explore, understand and then communicate relationships with the ocean further, and to do so with embodied methodologies (Sparkes and Smith, 2012; lisahunter and emerald, 2016a), I looked for ways to notice and capture engagements with the water; enfolding multi-perspective mobile space-time technologies, with body mapping and post-event elicitation. Anderson (2014) has used the frameworks of 'assemblage' and 'convergence' to convey surfing a wave, and a variety of other scholarship has focused on sensations of 'stoke', 'flow' or 'glide' as visceral surfing experiences of the wave (see, for example, Anderson, 2014; Stranger, 2010; Ford and Brown, 2006; Waitt and Frazer, 2012). Such work has attempted to provide new ways to methodologically and theoretically consider our visceral entanglements with water and make sense of surfing a wave, or going for a surf. And while there is increasing literature about surfing, and its sociocultural relationships, the dearth of work targeting the very act of surfing, in terms of the sensorial and situated specificity, particularly when going beyond the more-than-human ecological frame (and taking the very qualities of water into account) is notable.

Through several varied interactions – including surf lifesaving duties and competition, scuba-diving, helicopter surf rescue work, surfing, sailing, photography, researching and bodysurfing – I extend my previous work where I was attempting 'mindful spatio-sensational pedagogics to further my understanding of the sea/mind, as human, as self, and as not' (lisahunter, 2015b, 52). I return to surfing for this chapter's focus. Here I consider how an exploration of one's sensory experience, perception, sociality, knowing, knowledge, practice and culture may be informed by sensory autoethnography (Pink, 2015, 4–5). A sensory autoethnography is enfolded through this chapter; however, my primary focus is not to engage in a lengthy discussion of this approach, but rather on the outcomes of it, (re)presenting my field texts through a single surfing encounter using sensory narratives constructed via a range of technology (lisahunter and emerald, 2016b). The construction of field texts – I argue – provides a means to understand and communicate the practice of surfing. In what follows I discuss the methodological implications for 'knowing the sea', followed by onto-epistemological queries that might inspire action and symbiotic relationships that reduce harmful socio-environmental exploitation by humans all the while harnessing greater affordances of and for the sea.

A start point: constructing field texts

Sarah Pink describes a sensory autoethnography as 'a reflexive and experiential process through which academic and applied understanding, knowing and knowledge are produced' (2015, 4–5). As part of an ongoing ethnography associated with the broad field of surfing, I embarked on a series of sea-related encounters to pay attention to the senses. I constructed a range of what Clandinin and Connelly (2000) refer to as field texts (from direct experience), interim research texts (made in the moment of analysis, or what might be considered as autoethnographic narrative 'data'), and official publications and presentations. Building on earlier work (lisahunter, 2015b) my purpose here is to further experiment with the 'seaspacetimes' through surfing and to assess the reach of methodology and technology to understand and communicate this watery engagement. What kinds of field texts and interim research texts might be useful for making sense of the sea? How might these be communicated or *presented* to better understand or *represent* the experience? What 'technology' may assist in assembling these embodied moments, and in turn to better grasp and share the sensory and the more-than-human encounters of surfing? Below I share sensory narratives – field texts and interim texts – from one session of surfing. Field texts are retrospective writings made from and of direct experience (and are in plain font). Interim texts are my reflections at the time, often captured on video recordings and these offer 'in-the-moment' analysis (and are in italicised font). Following these textual (re)presentations, I move to an explicit discussion of methodology; of how such texts and the technologies used in constructing them may bring us closer in understanding and communicating how we live with the seas.

Stepping in: field texts and interim texts

1: Entering into experimental methods for thalassography

'I wonder what Donna Haraway (1991) would say to this?' (I chuckled). I walked down the deserted beach to the water's edge, adorned not just in the material culture of bikini top (worn for pragmatics and moral codes of changing from and to dry clothes in a public car park), but also a steamer (a full-length wetsuit used to keep warm), leaving only my feet, hands and head bare. Mine was a black suit with a surf company logo unsuccessfully coloured out by my permanent marker. I also had an array of cameras, one attached to the nose of the board, another fitted to a harness for my chest, another for my head and yet another to 'dangle' off the back of my board. I had a camera on the beach with a telephoto lens operated by a friend. All I had left to do was to finish dressing on the wet sand.

I put my board down, unravel the leash, roll up the wettie and wrap the leash around my back foot, secure it by rolling down my wettie again at the ankle. It all looks so serene. But looking through the land camera later I feel embarrassed and wonder at the feasibility of getting others to look like this and to engage with the sea using all those cameras, all that 'kit'. The selfie culture and assumed narcissism by others may limit who will don the gear and what sort of participant might be attracted to this sort of research.

My purpose was to see what slippery moments can be 'captured' and to understand if it is feasible to catch spoken thoughts and other senses through this technology while surfing, and to reflect on what senses are engaging with the surf. The purpose was to get closer to my relationship to the sea and vice versa.

It looked beautiful and the feel of just being out there was instantly good. I think, 'what do beautiful and good mean'? The sea was calm with small swells coming to greet the board-human-multiple camera complex as I walked out to sea. There was no one about, the air was cool to my face but the water on my feet was colder. I could feel the pressure of the wettie, tight across my chest, around my neck and on my forearms. There was a slight fishy smell or maybe a salty smell, or just a 'sea-ey' smell. The wave sounded as it broke, its colour shifting from a green to a tumult of white. My feet sank a bit further into the sand after each wave as though it was claiming me, beckoning me to get in. It felt soft against my skin, the tops of my feet were run over by the ... retracting, that's the word, retracting water (that word will do). 'I'd better get on with it before others start turning up' I thought. The front camera (on the board) was very off-putting. I was staring straight into it as I pushed off onto my guts and started paddling. 'I guess that's what the sea sees' (I laugh). 'Poor thing' (I laugh some more). I tied the other camera, tucked down my wettie for the paddle out, then to the leash once I was out there. I paddled over the first little broken wave. I hoped the camera would cope being dragged along and can pick up what's around.

I was paddling out further, sitting and waiting for waves, paddling but failing to catch waves and I was sitting out the back; just observing. This is what was prevalent in this surf session.

Perhaps I should focus on these moments, especially given most literature is about the 'caught wave'?

2: Sounding out the waves

'hmmmm (humming tune), hmmmmmmm, hmmmm, te ararau o Tangaroa, e rere kit e papaurunui, te ararau o Tangaroa, e rere kit e papaurunui, te ararau o Tangaroa, e rere kit e papaurunui, hmmmmmmm, hmmmmmmm, hmmmmmmmmm, whistling tune, te ararau o Tangaroa, e rere kit e papaurunui, te ararau o Tangaroa, e rere kit e papaurunui, te ararau o Tangaroa, e rere kit e papaurunui, hmmmmmmm, hmmmmmmm ... I was just 'here'. I wondered 'why that tune'? 'Why this humming'?

Looking back, I had caught myself humming a sound that I had not consciously begun. Was it the presence of the sea that inspired the song in the first place? The rhythm of the ocean reflected in the rhythm of the humming? Why not Beach Boys' 'surfer girl'? This was another one I'd been prone to catching myself humming or singing when no one else was around.

'I hope the sea is more forgiving than what a human audience might be if they were here' (I laugh)'

The video shows the sea almost still, I'm holding up the board looking out to the horizon, inviting a stroking of the surface by the hands and perhaps the tune stimulates the rhythm of the hands, or the rhythm of the water stimulating the tune and/or hands.

3: From not surfing the wave to surfing the wave

At one stage while facing out to sea I noticed the oncoming wave. A gentle set was approaching and I immediately saw my boardself on it, playing along with the surface of the wave, feeling the up and down motion and catching myself moving up and down. But I was not on that wave.

It struck me later when re-viewing that moment – followed by me and the board turning with the help of the ocean – that the surfing of the wave in some ways 'makes' the wave. This is not to say humans have total agency over the ocean. Rather they can be imbricated in it. It is neither just physical nor just human-constructed. It is both.

I was not on that wave. Rather, a gentle motion elicited images of the first diagram I had ever seen to explain the micro-circulation of water particles and the energy of the 'wave' passing through. I was a cork 'bobbing' until I spotted the next wave, and turning towards land with an unconscious flick of my legs in the opposite directions; descending my chest onto the board; digging one arm into the water followed by the other in a repetitive paddling motion, with my eyes watching over my shoulder to the path and supposed optimal energy point of the wave; speed from the water's pressure was against me and I slipped through the water and in a split second contraction and extension of my body, I was lifted forward and to the right, landward. I was no longer the cork but the result of the energy interacting with gravity as the face of the wave spilled forward, its convex shape giving way to concave. The wind was catching at my cap and hair, producing a 'coolness'. There was an 'exhilaration' of the ride. Fluid 'pumping in my legs'.

The wave pushes me along for a visual perspective; a perspective unattainable for non-surfers until early videography by George Greenough. The surf can now be 'experienced' by those who have never surfed and re-experienced by those who have surfed – through film/video, multisensorial and multimediated technologies of gaming and exergaming, virtual reality simulations and artificial wavepools.

4: (Im)possibilities of 'just surfing'

Since perhaps 2005 when I purposefully began to research ocean interfaces or perhaps even earlier when I was first taking the academy to the water, I realise that despite my best efforts it was extremely difficult, more difficult than I had imagined before entering the water, to 'just surf' in this context.

5: Meditation and the mundane, drifting and just being

Some time had passed and I was silent. I was just sitting there, gazing to the horizon just like I do in open-eye meditation, with the sea holding me up with a repetitive rise and fall. I know I switched off from thought but I was also very 'present'. I did think at some stage about the vibration; a sea of sensations. 'I' could feel the tiny breeze/cool on the left side of my face; sun warmth through the back of my wetsuit; coldness on my feet; the light pressure of the air and buoyancy of the water lifting my legs. My skin, pressured by the wetsuit, felt more intense in my feet, hands and face with the mix of temperature and non-wetsuit pressures. I felt these sensations but they were also somehow further out than my body, mixing with air, water and board.

I know there is talk of 'convergence' in the wave literature (Anderson, 2014) *but what about 'getting lost' in the everyday moment of and waiting for waves? What of the presence of yet 'others' – fish, sticks, seaweed, technology? What of the emptiness experienced by some as peace and solitude and by others as isolation, loneliness and/or despair* (Peters, 2012)?

It felt like a fullness. It felt like being part of something so much bigger. I was connected, much like those who express the sublime (Anderson, 2014) but it also had shadows of insecurity (don't know what was going on there). Perhaps it was fed by the sensationalisation of shark attacks in the media and my recent viewing of *Soul Surfer* based on the experiences of a well-known surfer, Bethany Hamilton, and her shark experience and loss of her arm.

The AV had recorded a muffled sound that I remember to be what I thought was an internal laugh at the paradox of beauty and peace with danger and tumult – affects triggered by intertextuality and memories rather than the present. I think I rationalised that the sea is home to more creatures than I can comprehend and I'm just a visitor, hopefully one with some awareness about the nature of the sea, its habits and inhabitants.

Stepping away: reflections on approach

As the texts and interim texts reveal, the moment of analysis – of understanding and communicating – already began in the consciousness/articulation/commentary following the experience and in the reflections on the video recordings. But to tease out 'analysis' in this very messy, sensual, multi-timespace project, I paid attention, perhaps imperfectly, to what was 'jumping out' of the assemblage of texts. This is what was presented above, with analysis continuing not just after textual construction but possibly before and as part of an ongoing act of 'making sense'. There is a lack of certainty and fixity in the texts. They are (hopefully) not always easy to follow, or to understand. In this sense, they succeed in embracing the difficulty in even understanding my own engagements. They are full of fluidity and enfolding. But they (re)present a moment of 'being' and 'living' with the sea. In this 'being in' and 'looking back' through text and video, in the moment and in the moments after, I was able to take stock of what is it to engage with the sea. In surfing, it is not all in the wave. The wave is not the only articulation of the sea–space–time assemblage. It is the waiting. The paddling. The *not* surfing.

Pragmatically this exploration was aimed at identifying whether there was more to capture through an embodied, creative approach and to reflect on what such capturing and then articulation might do to promote 'better' understanding. Such an embodied approach

> validates the need to understand how bodies feel internally – sensations, moods and physical states of being – in relation with material social space. The relational aspect of the concept is substantial, as it has been successfully used to merge seemingly individual concerns of the experiential/biological body with cultural blueprints, social orders and economic structures.
>
> (Sweet and Escalantes, 2014, 3)

In constructing my field texts I discovered new realisations that I had not had whilst in the moment, or 'in situ'. The different seaspacetimes provided by constructing the texts – field and interim – whilst 'looking back' proved valuable for reflection and bringing to the surface some of the sensory experiences I had, and continued to have, as a result of the surf and of the research.

Researchers, some inspired by the 'narrative turn', have engaged with 'story-ing' techniques in different ways and at different points within the research process in the production of field texts, as analytic methods to create interim texts and as research texts for dissemination/communication of the 'so what' or 'what now' of research (Clandinin and Connelly, 2000). As I have shown here, narrative as a form of inquiry and/or (re)presentation in the collation and curation of field and interim texts may capture and communicate the embodied, emplaced and multisensorial experiences of physical culture. Our storied worlds are, by their very nature, embodied, emplaced and multisensorial. However, to make sense of participants' experiences of their lived worlds, we are often forced to reduce these stories to words on paper, as flat interview transcriptions, ordered field observations, coded and remodelled summaries or fragments of narrative written by the participants or researchers. As such, although Dewey's (1938) classic work on experience has progressively been taken up by the social sciences, it is yet to be fully operationalised. Is there a way of better articulating sensory experience?

Limited as scholars are by written text, our understandings of experiences in physical culture, embodiment and movement faces the ongoing challenges of capturing embodied/sense experience as narrative and analysing, representing and even reconstructing storied worlds in embodied ways and as (embodied) knowing. I acknowledge there are many limitations to the field and interim research texts above as 2-D (re)presentations – but we can use strategies of evocation to provoke affects in those who read them. We can use visual cues and other sensory tools (relating to sound or smell) to make *textual differences* that more fully evoke the multiple seaspacetime encounters in the research text construction.

The extent to which this narrative can register some of the unconscious material remains to be seen. What, as a reader, for example, did you make of field text and interim text 5 and how much could you relate to my 'being in the present with the sea?' But the point is, such articulations are active, even if they remain partial. Two-dimensional or flat texts do 'get' humans to connect with a message because they are affective. And provocatively constructed they can be employed affectively to communicate and assist in garnering understanding. If we can get *closer* to the water through such techniques it may help not just the researcher but those to whom the research is disseminated to understand a raft of sea-going political relationships: global ownership, exploitation, pollution, surf rage, surf activism.

Indeed, while relatively aware of the issues that arise when humans interact with the seas, my methodological exploration nevertheless deepened my own awareness. I re-evaluated my ecological commitment, my negotiation of limits, my justifications for acts and behaviours and the potential deleterious effects that I might perpetuate for the materials I use, and the practices I buy in to. Yvon Chouniard, founder of surf company Patagonia, has openly criticised those in the surfing industry – surfboard makers, surf clothing manufacturers and surfers more generally – for 'their lack of environmental leadership and independent thought' (Latourrette, 2004, cited in Hill and Abbott, 2009, 292). Like Hill and Abbott, and others doing similar work, my aim has never been 'to demonize the culture, but rather to present a critical consideration at surfers, and surf culture's place in the environment' (Hill and Abbott, 2009, 292). This has been through the critical lenses I have employed in previous work (lisahunter, 2015a) but also now, as a result of the exercise articulated here. Stimulated by the study and its write-up – of sensory engagement and deep political ecology – I have become conscious of the more-than-human assemblage involved in the act of surfing through reflection on the conscious and also less-than-conscious moments: the being in the sea, its movement, the sea life to which it is home, the sounds, the smell. It is important to continue to be conscious of the effects of surfing as an ongoing commodified, neo-colonial project (lisahunter, 2017, 2018a, 2018b, 2018c). It is important to be conscious of the effects of any engagement with the sea: from walking on its edges, to diving in the deep.

Stepping up: embracing emerging methodologies

Methodologies that attempt to engage with new materialism (see for example Lovino and Oppermann, 2014; Somerville, 2016) and a 'wet ontology' (Steinberg and Peters, 2015) – the very livingness of our world – may well require a more novel set of approaches to make sense of and communicate that world. Echoing feminist work from the last century that has fleshed out the materiality, carnality and sociality of the body (for example, see Gatens, 1988; Grosz, 1994; Grosz and Probyn, 1995), Sarah Whatmore has described and encouraged a return to materiality in what she saw as the 'most important aspects of an ongoing realignment of intellectual energies' (2006, 604). Her efforts have been to work towards a bringing together of the 'bio' and the 'geo'; human life and the living environment (Whatmore, 2006, 604) in a theoretical (re)turn towards worldly engagement. As part of the (re)turn she has also suggested focusing on practice, affect and the politics of knowledge to explore such co-fabrication between human and more-than-human worlds. Pink (2011, 2013, 2014) too has promoted similar outcomes of focusing on practice and more-than-human worlds with the development of ethnography, initially visual but then also sensory and digital.

Yet sensory accounts of human experiences are typically embedded in the more-than-human *terrestrial* spacetimes. In the case of the surfed wave, in the case of *watery worlds*, we need more sophisticated methods of creating sensory accounts of fluid spaces, contemplating how the bodies move through and sense these spaces. An epistemology of the senses has mediated the more popular moments of surfing engagement – the 'stoke' and 'flow' – in surfing. Stranger describes being 'stoked'; the 'high' or thrill that surfing can induce, the 'ecstatic reaction to a surfing session or a particular ride' (Stranger, 1999, 275). Anderson suggests assemblage and convergence, theorised as the 'flow' experience to understanding the moment a surfer 'lives' the wave and the conjoining of body, suit, wave, velocity (2014, 76). There is space for other affects to be explored (lisahunter, 2015b), including being 'wiped out over the falls' (what happens when the board-human tumbles or crashes from the lip of the wave at the top down in front of the wave path; it is generally regarded as an unpleasant experience and, at worst, can be life-threatening), or injured by another board-human or sea creature (such as sharks, popularly demonised in popular discourse, see Gibbs and Warren, 2014), and, as I have reflected on here, the more banal or mundane but nonetheless important moments of surfing: the bobbing, sitting, waiting, reflecting. In this space in particular, I have recognised the seaspacetime, for me at least, of thinking beyond the surf itself and rather the very nature of my engagement with water, and in turn my ecological commitments to the sea.

Personal attached waterproof devices, Point of View devices (POV), drones, and even the land-based camera reveal different things from different perspectives; each, also with different limitations (see also Spinney, 2009; Spinney, 2015) in relation to capturing cycling experience from multiple perspectives). The chest mount contributed to greater sacral pain but showed forward of body images missed by the headcam. Whether what was missed had anything to add to the headcam remains to be 'seen'. Drones were out of my budget but their innovative 'aerial' view may give a different visual 'picture' of a given scene (this is evident when I watch footage of sea creatures close to unaware surfers, captured from the skies when carrying out surf patrol in a helicopter). Evers (2015) has also played with technology and the surfing experience, holding a GoPro in his mouth. I felt uncomfortable reading his experience, imagining what a mouth-held camera would do to my breathing and commentary. The devices also become part of the assemblage and affects of surfing. As relayed earlier in the chapter, I felt embarrassment when initially looking at the board-mounted camera, face-on, and imagining if anyone else came to the scene and witnessed such a 'cyborg'. The more-than-human interface, when such technology is used, is not just about entanglements with a fluid watery environment but also an entanglement in technology. This has onto-epistemological implications as well as methodological questions where the material effects of the technology may change the experience and our way of experiencing it physically, socially and perhaps sensorially (see Pink et al., 2016). But where the use of technology, especially video and cameras, is becoming commonplace, its impacts may not be so profound. Already in other sporting scenes there is a proliferation of cameras, including Point of View (POV), cameras which is garnering the attention of scholars (see for example Hutchins and Rowe, 2013). The so-called 'selfie' culture cultivates board-mounted cameras to capture the surfer rather than the ride-board-surfer-wave in the eye of another. Such technology easily and (relatively) cheaply captures personal facial expressions, voice, fine body gestures and the world beyond the surfer regardless of the direction faced.

Leszczynski discusses spatial media/tion to consider how 'the pervasiveness of more recent technologies necessitates an epistemological entry point from which to engage with spatial media as artifacts, materialities, presences, and sites of relationality that not only broker and disrupt, but also actively constitute, our "encounter[s] of self and world"' (2015, 731). As a means of knowing, digital media has been ubiquitous and it is no longer unheard of to have a camera passing by, out the back, or to see amateurs filming each other or themselves surfing. Nor is it unusual, as previously stated, to see evidence of surfboard deck-mounts for cameras, or surfers with wrist cams and selfie sticks. Such media

> provides an entry point for considering mediation itself as an ontological claim about the nature of our everyday being-with-each-other, and our being-with location-based objects and services that 'help to constitute what is real for us, and what we are in relation to that reality'.
>
> (Verbeek, 2011, 392)

'Reality', of course, may be understood as already always mediated, 'the product of the myriad intersections and mutual constitution of technology, society, and space relations that are themselves the products, or effects, of mediation' (Leszczynski, 2015, 3). As such, seaspacetime of 'the surf' (as an individually surfed wave or the whole session's components) are a convergence/assemblage (as Anderson, 2014, puts it) not just of the aquatic, human, board relations but also technology. But how might these technologies be used for enviro-socially positive and sustainable outcomes, as well as new possibilities for being/sensing and interacting? How might they engender a sensitivity to location, user relations and elements affording, as Moores noted an 'instantaneous pluralisation of place' (2012, 16)?

More-than-human watery connections through sensory narrative

As the sea let me go and left me, the beach took my weight and I pressed upon its dry sand. I could easily stop and create a coherent narrative about the waves that took me/I took; the conditions that enabled 'a surfed wave'; and – as I have posited here, the nonetheless productive conditions that did not.

Conscious debriefs, whether written down, or not, are common when the direct sea–human relationship is over, or even when humans are still in it, working with moments and memory to create a story from the last wave or the last surf, in other words to make sense of it, to understand. But my own surf encounters – as the focus of my research interests – has pushed me further to try and articulate and communicate such encounters and to do so in ways that better capture that slippery, embodied, visceral and worldly experience. But this has had ontological, epistemological, methodological, theoretical implications.

This chapter has (re)presented one single surf session. The field texts are, inevitably, *constructs* (a point from which I have not shied away – understanding and communication is never just natural; it is made and formed). Experience has been constructed cognitively, analysed and then articulated. But, as I have also noted, this is not without the agency of water and non- and more-than-human actors. Without the sea this chapter would not *be*. Just as the 'surf' 'the surfed wave' and 'the surf session' are co-constructed, so too is this chapter. It is formed and shaped by my own writing, by feedback, by editorial comments. It is not a 'pure', 'exact' account. Even you, the reader, have now folded into the text and will read 'the experience' through your own eyes. A text is shaped by multiple timings of memory – experience, telling/recording, reflection, analysis, re-reflection, re-recording, editing, honing and a 'final' reconstruction.

Through this construction, built through a variety of methods, I have empha-sised the need to pay attention to experiences evoked through being *in* the sea, where:

> a new type of relational place is experienced through becoming part of the surfed wave. Coupled to modern metaphors, historical technologies, and scientific debates, surfers' perspective of the sea can therefore supplement and enhance our attempts to chart new maps of water worlds. Gaining insight from individuals' experiences of the place of the surfed wave also emphasizes the importance of corporeal engagement with the sea.
>
> (Anderson, 2014, 83)

However, I would argue that realising the experience, affect and relationship with the sea per se, as well as just 'the wave', is crucial. The mundane waiting, the meditative spaces, the return 'out the back' or to the shore after a wave needs further consideration, if we are to understand the heterogeneity in surfing experiences, the varied assemblages that constitute surfing and the ways we, as humans, are imbricated in water worlds. Whether the wave is the 'anchor' for surfers' lives, as Anderson (2014) suggests, or whether it is other components, there is still much more we can explore ontologically, epistemologically, theore-tically and methodologically. But why do so?

To my mind there is no doubt that paying attention to the sea is vital (Anderson and Peters, 2014), as an embodied memory constituted materially and phenomenologically. I would also agree with Orr (1992, 146) that '[t]he symptoms of environmental deterioration are in the domain of the natural sciences, but the causes lie in the realm of the social sciences and humanities'. While I have not yet returned to the other contexts and elements that co-created the field texts here (terrestrial experience and so on), there is a wealth of lessons we can take from engaging with, reflecting upon, articulating and communicating watery experience and in understanding how 'matter' comes to matter to human experience (Barad, 2003). It is this engagement that may assist in developing a strong sense of environmental ethics, as I have suggested.

And, whilst agreeing with those who encourage a 'wet ontology' (Steinberg and Peters, 2015) and more-than-human approaches to practices such as 'the surf' (Anderson, 2014), we need also to be careful not to set a 'solid/liquid' binary. Rather, we need to work with 'states' that are in constant flux by dint of being atoms (Steinberg and Peters, 2015). As Steinberg and Peters note, 'waves, and the wet ontology they exemplify, may be exceptionally well suited for understanding the politics of our watery planet' (2015, 261). This understanding and its communication not only requires new methodologies but, in partnership with them, new onto-epistemologies. I would add that waves, individually, as a 'session' assembled with and beyond the human, and beyond the human encounter but nevertheless entangled in it, are key areas for ongoing sensorial exploration if we are to ensure a more positive future for the oceans and therefore ourselves. I argue that this is crucial to inspire action and symbiotic relationships that reduce harmful exploitation by humans all the while harnessing greater affordances of, with and for the sea.

References

Amadio, N. (1993). *Pacifica: Myth, Magic and Traditional Wisdom from The South Sea Islands*. Sydney: Angus and Robertson.

Anderson, J. (2009). Transient convergence and relational sensibility: Beyond the modern constitution of nature. *Emotion, Space, and Society*, 2, 120–127.

Anderson, J. (2014). Merging with the medium? Knowing the place of the surfed wave. In J. Anderson and K. Peters (Eds.), *Water Worlds: Human Geographies of the Ocean*. Farnham: Ashgate, pp. 73–85.

Anderson, J., and Peters, K. (Eds.). (2014). *Water Worlds: Human Geographies of the Ocean*. Farnham: Ashgate.

Barad, K. (2003). Posthumanist performativity: Toward an understanding of how matter comes to matter. *Signs: Journal of Women in Culture and Society*, 28(3), 801–831.

Brown, M., and Humberstone, B. (Eds.). (2015). *Seascapes: Shaped by the Sea. Embodied Narratives and Fluid Geographies*. Farnham: Ashgate.

Clandinin, D. J., and Connelly, F. (2000). *Narrative Inquiry: Experience and Story in Qualitative Research*. San Francisco: Jossey-Bass Inc.

Dewey, J. (1938). *Experience and Education*. New York: Macmillan.

Evers, C. (2015). Researching action sport with a GoPro™ camera: An embodied and emotional mobile video tale of the sea, masculinity and men-who-surf. In I. Wellard (Ed.), *Researching Embodied Sport: Exploring Movement Cultures*. Abingdon: Routledge, pp. 145–163.

Ford, N., and Brown, D. (2006). *Surfing and Social Theory: Experience, Embodiment, and Narrative of the Dream Glide*. London: Routledge.

Gatens, M. (1988). Towards a feminist philosophy of the body. In B. Caine, E. Grosz and M. De Lepervances (Eds.), *Crossing Boundaries: Feminism and the Critique of Knowledges*. Sydney: Allen and Unwin, pp. 59–70.

Gibbs, L., and Warren, A. (2014). Killing sharks: Cultures and politics of encounter and the sea. *Australian Geographer*, 45(2), 101–107.

Grosz, E. (1994). *Volatile Bodies: Toward a Corporeal Feminism*. St. Leonards: Allen and Unwin.

Grosz, E., and Probyn, E. (1995). *Sexy Bodies: The Strange Carnalities of Feminism*. London: Routledge.

Haraway, D. (1991). *Simians, Cyborgs, and Women: The Reinvention of Nature*. New York: Routledge.

Hill, L., and Abbott, A. (2009). Surfacing tension: Toward a political ecological critique of surfing representations. *Geography Compass*, 3(1), 275–296. doi:10.1111/j.1749-8198.2008.00192.x

Hutchins, B., and Rowe, E. (Eds.). (2013). *Digital Media Sport: Technology and Power in the Network Society*. Abingdon: Routledge.

Leszczynski, A. (2015). Spatial media/tion. *Progress in Human Geography*, 39(6), 729–751.

lisahunter. (2015a). 'Stop': 'No'. Exploring social suffering in practices of surfing as opportunities for change. In lisahunter W. Smith and e. emerald (Eds.), *Pierre Bourdieu and Physical Capital*. Abingdon: Routledge, pp. 47–56.

lisahunter. (2015b) Seaspaces: Surfing the sea as pedagogy of self. In M. Brown and B. Humberstone (Eds.), *Seascapes: Shaped by the Sea. Embodied Narratives and Fluid Geographies*. Farnham: Ashgate, pp. 41–54.

lisahunter. (2017) Desexing surfing? (Queer) Pedagogies of possibility. In D. Zavalza Hough-Snee and A. Eastman (Eds.), *The Critical Surf Studies Reader*. Durham: Duke University Press, pp. 263–283.

lisahunter. (2018a). Queering surfing from its heteronormative malaise: Public visual pedagogy of circa 2014. In lisahunter (Ed.), *Surfing, Sexes, Genders and Sexualities*. Abingdon: Routledge.

lisahunter. (2018b). (Counter)cultural changes in surfing and surfing scholarship: Towards diverse, intersectional, liminal, complex and queer activism dialogues. In lisahunter (Ed.), *Surfing, Sexes, Genders and Sexualities*. Abingdon: Routledge.

lisahunter. (2018c). The long and short of (performance) surfing: Tightening patriarchal threads in boardshorts and bikinis. *Sport in Society*, 1–18, (early online). DOI: 10.1080/17430437.2017.1388789.

lisahunter and emerald, E. (2016a). Sensory narratives: Capturing embodiment in narratives of movement, sport, leisure and health. *Sport, Education and Society*, 21(1), 28–46.

lisahunter and emerald, E.. (2016b). Sensual, sensory and sensational narratives. In R. Dwyer and I. Davis (Eds.), *Narrative Research in Practice: Stories from the Field*. Dordrecht, The Netherlands: Springer, pp. 141–157.

Lovino, S., and Oppermann, S. (Eds.). (2014). *Material Ecocriticism*. Bloomington: Indiana University Press.

Moores, S. (2012). *Media, Place and Mobility*. Basingstoke: Palgrave Macmillan.

Orr, D. (1992). *Ecological Literacy*. Albany: State University of New York Press.

Peters, K. (2012). Manipulating material hydro-worlds: Rethinking human and more-than-human relationality through offshore radio piracy. *Environment and Planning A*, 44, 1241–1254.

Peters, K. (2015). Drifting towards mobilities at sea. *Transactions of the Institute of British Geographers*, 40, 262–272.

Pink, S. (2011). Images, senses and applications: Engaging visual anthropology. *Visual Anthropology*, 24(5), 437–454.

Pink, S. (2013). Engaging the senses in ethnographic practice. *The Senses and Society*, 8(3), 261–267.

Pink, S. (2014). Digital–Visual–Sensory design anthropology: Ethnography, imagination and intervention. *Arts and Humanities in Higher Education*, 13(4), 412–427.

Pink, S. (2015). *Doing Sensory Ethnography* (2nd edition). London: Sage.

Pink, S., Horst, H., Postill, J., Hjorth, L., Lewis, T., and Jo, T. (2016). *Digital Ethnography: Principles and Practice*. London: Sage.

Somerville, M. (2016). The post-human I: Encountering 'data' in new materialism. *International Journal of Qualitative Studies in Education*, 29(9), 1161–1172.

Sparkes, A., and Smith, B. (2012). Embodied research methodologies and seeking the senses in sport and physical culture: A fleshing out of problems and possibilities. In K. Young and M. Atkinson. (Eds.), *Qualitative Research on Sport and Physical Culture*. Bingley: Emerald, pp. 167–190.

Spinney, J. (2009). Cycling the city: Movement, meaning and method. *Geography Compass*, 3(2), 817–835.

Spinney, J. (2015). Close encounters? Mobile methods, (post) phenomenology and affect. *Cultural Geographies*, 22(2), 231–246.

Steinberg, P. (2001). *The Social Construction of the Ocean*. Cambridge: Cambridge University Press.

Steinberg, P. (2014). Foreword on thalassography. In J. Anderson and K. Peters (Eds.), *Water Worlds: Human Geographies of the Ocean*. Farnham: Ashgate, pp. xiii–vii.

Steinberg, P., and Peters, K. (2015). Wet ontologies, fluid spaces: Giving depth to volume through oceanic thinking. *Environment and Planning D: Society and Space*, 33, 247–264.

Stranger, M. (1999). The aesthetics of risk: A study of surfing. *International Review for the Sociology of Sport*, 34(3), 265–276.

Stranger, M. (2010). Surface and substructure: Beneath surfing's commodified surface. *Sport in Society: Cultures, Commerce, Media, Politics*, 13(7), 1117–1134.

Sweet, E., and Escalante, S. (2014). Bringing bodies into planning: Visceral methods, fear and gender violence. *Urban Studies*, 52(10), 1826–1845.

Verbeek, P. (2011). Expanding mediation theory. *Foundations of Science*, 17, 391–395.

Waitt, G., and Frazer, R. (2012). 'The vibe' and 'the glide': Surfing through the voices of longboarders. *Journal of Australian Studies*, 36(3), 327–343.

Whatmore, S. (2006). Materialist returns: Practising cultural geography in and for a more-than-human world. *Cultural Geographies*, 13, 600–609.

8 Science and culture

Transitioning currents in times of climate change

Susan Reid

The incredibly beautiful and, for some, frightening ocean is likely to be the largest moving 'thing' most of us will ever witness. Ocean waters tumble and cascade down ridges, creep abyssal plains, pinch off in crisp eddy rings that coalesce at the surface, rise up in steep shoals and upwellings, crash and fold in on themselves and evaporate to the clouds. Ocean currents[1] are not stable or infinite – they have histories and may be shaken, diverted and propelled at different speeds and in different directions by planetary wobbles, winds, celestial forces and Earth's orbit. Through their global journeys and local upwellings, salty, wet, transitional currents animate the ocean's complex liveliness. Enormous circulatory systems glide, wind and pump their way around the earth, crossing abyssal plains, massaging continents, breaking off into meandering currents and eddies that deliver warmth and nutrients before cycling back to be replenished.

Imagining the world through currents takes us offshore, away from terrestrial biases to consider our relations with ocean life and dynamics. Driven by transitional material exchanges, the surface currents, overturning circulation and biological pumps influence planetary climate systems and biological communities through myriad relations. They are thoroughly implicated and entangled with terrestrial events and intensifying human activities: fluxes of carbon, plastics and other materials arriving, transforming and leaving through river mouths, canal outlets, vehicle exhausts, rain-freshened surfaces, holiday coastlines, beneath ice sheets and evaporating skyward. Currents send ashore the material missives of the local conditions through which they pass: washed up wrack, froth and spume from offshore sediment and kelp forests; as they also send warmth, seeds, remnant oil and plastic from one basin edge to another. As Neimanis (2014, 19) observes of watery bodies, the distributive capacity of currents is both situated and global. Currents also stream material transitions, showing us where change comes from and where the effects of change are going.

The proposition of 'living with the seas' is framed companionably and suggests a relationship of planetary neighbourliness. For despite destructive tsunamis and powerful waves, the ocean[2] is a generous provider of life conditions for humans and other animals. However, exploitation of marine environments swirls in the undertow and challenges the very concept of companionability. In an unequal exchange, the ocean's biological communities, environments and dynamic systems are imperilled by declining fish stocks (Pauly, 2010; Probyn, 2016); loss of marine biodiversity (Worm et al., 2006; Mitchell, 2009); sand depletions (UNEP, 2014); pollution from offshore petroleum extraction (Passow, 2015); threats of deep seabed mining (Hunter and Taylor, 2013; Rosenbaum and Grey, 2015); and global material flows of plastics that now speckle planetary currents (Cressey, 2016). Intensifying streams of our carbon emissions intersect with these issues and bring transition of an entirely different register. Planetary scaled ocean warming is deepening (De Lavergne et al., 2014; Hill, 2013), and this is significant for the flow and reach of ocean dynamics and the creatures and other systems reliant on them (Van Gennip et al., 2017; Brierley and Kingsford, 2009; Eissa and Zaki, 2011). We humans are in a soak with the ocean – dependent on the very systems and materials that our actions imperil; and thoroughly implicated in, and affected by, its transitions.

Governance systems have failed to remove these ongoing threats. The ocean that informed the long drafting of the United Nations Convention on the Law of the Sea (UNCLOS) is not the same ocean today, yet its legal representations remain. Particular provisions in governance regimes also amplify the ocean's vulnerable state. For example, UNCLOS provides the legal basis for ocean exploitation, including seabed mining – an industry that may destroy some of the least understood ecosystems on the planet (Earle, 2016; International Seabed Authority, 2011). Legal theorists Grear (2015) and Bosselmann (2010) rightly attribute the enabling of environmental exploitation to reductive discursive strategies deployed by law. Guided by values still alive to its Enlightenment provenance, UNCLOS puts nature to work in the service of human enterprise.

Under the extractivist regimes in which we are all differently implicated, conservation measures and principles may mitigate against the environmental harms of these activities. However, they have not yet prevented ocean transitions that augur less comfortable futures for differently vulnerable human and other animal communities. Foundational values underpinning UNCLOS instrumentalise a world for our use; one that is exterior to, and situated outside, the world of relations, rather than a world imagined from within its makings. The spatialised and territorialised realm that is the law's ocean is structured as a background place where things happen – fish are harvested, sediments are mined, cables are laid, bombs are tested, chemicals are dumped, plastics circulate, waters warm and acidify (see also Steinberg, 2001). In this imaginary the ocean is instrumentalised as quarry, pantry, sink and sump.

Sea truthing through transitioning currents

How we understand nature and the environment, as Neimanis et al. remind us, has 'implications for laws, policies and individual actions' (2015, 482). See also Code, 2006; Castoriadis, 1987; Plumwood, 2002). Imaginaries are both crucial to environmental governance (Neimanis et al., 2015, 482), and to legitimising governance regimes (Steinberg, 2001). This chapter does not offer a critique of governance regimes, rather it encourages us to think with ocean currents to find conceptual openings for new, ecologically attentive imaginaries that might enable us to contemplate how we might dwell more companionably in relation with the ocean. The environmental justice focus and ecological orientations that unfold here diverge from the approach of other authors who use the ocean as a medium through which to develop theory. See, for example, critical geography theorists Steinberg and Peters' (2015) rethinking of social and political issues through the conceptual frame of wet ontology, and Stefan Helmreich's (2009) work with the ocean's lively microbial world.

I propose that to 'sea truth' and to live better with the ocean, we might understand more closely the material agency and transitional nature of the ocean's dynamic systems and relations. Doing so offers a conceptual ground for understanding the nature of transition itself, so that finding ways to live better with the ocean might challenge us to live better with its transitions. Beyond their narrow taxonomic categories, or as subjects of siloed scientific research, how do ocean currents and other moving bodies shape the liveable worlds that humans and other animals experience, albeit differently, across different localities. What might ecologically oriented imaginaries tell us of how our material appetites and carbon flows impact the ocean? How might we create imaginaries for more companionable ways to live well with the transitioning ocean in times of climate change?

The chapter's focus on ecological thinking, relationality and transition brings into view some of the ways that materialities enliven the ocean and implicate its transitional nature. I am informed by insights from new materialisms and posthumanist feminisms, such as Lorraine Code's ecological thinking framework which approaches knowledge-making in a way that is 'sensitive to human, ecological, historico-geological diversities' (Code, 2012, 86). Extending Code's approach into the ocean, I work affectively and imaginatively across marine oceanographic texts, which I am reliant on for insights into the remote, deep ocean. Close reading and imaginative engagement with these texts finds opportunities to synthesise knowledge across disciplines. In the process of thinking with the ocean, mediated through human epistemes, I draw on aesthetics and lyrical theoretical approaches as valid, critical tools for discerning unseen connections and implications of ocean relations and transitions.

In the first part of this chapter I explore some of the ways that currents move about the ocean, transitioning with the material relations of nature/culture entanglements. Then I offer two short meditations that provide speculative perspectives and potential imaginaries on the nature of transitions associated with very different dynamic systems: the biological pump that is created by the collective efforts of Southern Ocean krill; and the Antarctic Circumpolar current. The chapter closes with a short meditation on the nature of transition itself.

Basin relations – relational currents on the move

Surface currents and winds work together to up-end our terrestrial assurances of place. Currents are described by the direction in which they move, whereas the winds that drive and buff them are described according to the direction from which they come. Powerful westerly trade winds chase the easterly flowing Antarctic Circumpolar Current (ACC). Meanwhile, the East Australian Current (EAC) hugs the continent's east while travelling the western boundary of the South Pacific Basin. The Leeuwin Current tracks along the west coast, travelling the eastern boundary of the Indian Ocean. Brushing, swirling, licking and sweeping the continents and islands, and their submarine shelf slopes, currents animate and transform the solid edges against which we orient ourselves: seaward of the eastern coast of Australia, for example, a concert of uniquely constituted rotating water masses, chains of eddy lenses and coalescences of cast-off rings stir and unsettle the western edge of the Pacific Basin.

Currents are not undifferentiated – in an ecological sense, each has its own histories, watery routes, marine passengers, physical characteristics and ways of responding and transitioning with the material conditions they meet. Nor are current flows assured or immutable, as Scher et al. (2015) have found in relation to the ancient westerly flowing origins of the ACC. Stefan Helmreich (2014), too, explores the physical variations of waves but with a focus more on their cultural significance. Currents have their own way of going about the ocean, meandering snug against rocky outcrops stubbled with crimson anemones, refreshing frilly green algae, or streaming through sediment plumes that rise from seabed excavations.

Transitioning currents and carbon streams

Currents act and can be acted upon with an inherently relational and ethical agency; subtly transforming the waters through which they pass, and in turn transformed by the waters that they meet, the particles collected, and the continents and seafloors they sweep. Karen Barad's concept of 'intra-active' describes matter's lively and co-generative agency and lends itself well to ocean relations (Barad, 2007). Human actions too materially transition with ocean currents, in the sense that Barad refers to as 'intra-acting from within and as part of the world in its becoming' (Barad, 2007, 396). Sticky, salt watery currents manifest ocean relationality, and carry our weathering agency, and streams of carbon and plastics, into thick time. If '[w]ater is eminently natureculture' (Neimanis, 2014, 15), then so too is its myriad expressions as ocean currents. Temporalities and materialities of 'natureculture' (see Donna Haraway's *The Companion Species Manifesto* (2003)) entangle currents: emissions and engine ignitions, wireless laptops, residential dwelling, bridges, chemicals and minerals, atmosphere and ocean skin, heat and volition – all co-producing the signal changes now coming into visibility. Transitioning natureculture relations of ocean currents carry plastics from river mouths, and gather them in the grind of the North Pacific's *Great Pacific Garbage Patch*, or as micro-plastics ingested by plankton. They also observably catch up with us in the Southern Ocean, where around 60 per cent of anthropogenic heat and almost half our carbon dioxide emissions are absorbed (De Lavergne et al., 2014). Carbon emissions and heat penetrate deepening levels of warming surface waters. The consequence of these transitions now measurably influences the distribution of temperature into ancient bottom waters and overturning – natureculture transitions carried in the long-term memory of ocean circulations. As watery bodies, currents convey what Neimanis (2013, 32) describes as the watery archives of human consumption.

Emerging evidence of the thermohaline circulation's transition through natureculture material entanglements signals a need for a more ethical approach to ocean relations. The long-range temporal scale of such systems speaks little of their vulnerability to changing material intensities. Its planetary scale belies the exquisite exchange and transitioning of materials at the heart of its movement. One stage of this transition occurs when ice forms around the polar region and, in the process, it spits salt back into the sea. This dense, cold, saline water sinks and, as it does, water moves in to replace the sinking water and eventually a current is created. Cold deep water moves horizontally across the seabed until it can rise again to the surface, usually around the equatorial regions. In the thick time of the thermohaline overturning, its watery archives move at the rate of a few centimetres per second. Some ancient water can take up to 1,000 years to complete the cycle, and as many years to see the sun (Mann and Lazier, 1996).

Equally planetary in scale, carbon and other greenhouse gases thicken the atmosphere and the ensuing warmer ocean sets in train a different type of transition. As carbon-plump surface waters melt polar ice, an excess of fresh water is poured into the thermohaline mix. These diluted polar waters enfeeble the thermohaline exchange to the extent that there may be real potential for a thermohaline circulation collapse (Keller et al., 2000, 19). Ecological knock-on effects of climate change are also transitioning the East Australian Current (EAC) and profoundly changing ocean relations. Unsuited to the warming waters, marine creatures such as yellow-tailed kingfish and coral trout retreat further south (Milman, 2015). Spiny sea urchins have arrived in Tasmania and are devouring underwater forests of the 30 metre tall giant kelp, Macrosystis pyrifera (van Sebille et al., 2014). Tasmania's cool waters creatures are caught unprepared by the arrival of new warm water predators (Kelly, 2011). As van Sebille et al. (2014) note, these creatures can only move so far before they meet the edges of the continental shelf – this is the end of the line as their next habitable shelf is about 3,000km south at Antarctica.

Stacy Alaimo's concept of transcorporeality aptly captures the movement of materials across bodies and highlights the porosity of boundaries between them (Alaimo, 2010). 'Body' is broadly conceived in this chapter to encompass human bodies, other biological bodies and creatures, ocean currents, eddies and pumps and molecules of organic and inorganic matter. Using 'bodies' in this fluid sense also emphasises the movement of materials across porous boundaries. Transcorporeal movement of ocean bodies can be understood in the way currents travel and trouble the edges of marine park delineations, abstract jurisdictions and mining tenements. Currents animate the flow of life between and through boundaries, carrying marine creatures into invisibility or the vulnerability that comes with dwelling in the spaces between. Currents insist that boundaries will be breached.

Transcorporeality also emphasises shared or differently experienced physical vulnerabilities across human and more-than-human worlds. Such relational material exchanges remind us that bodies are always 'becoming in webs of mutual implication' (Neimanis, 2013, 25). Excess carbon molecules that move from the atmosphere into ocean waters, for example, transform the nature of the water by increasing its temperature and/or acidity. Not only do such intra-active material exchanges have the potential to transform the materials being exchanged, but the nature of relations between bodies also transitions.

Ocean currents are exemplars, par excellence, of intra-active transitional relations – the process which recognises that as bodies transition, so too do their relations. Currents transition by carrying and subsuming warmth, carbon, plastic and other material expressions of humannature. As they do, their relations with other bodies also transition. For example, as the East Australian Current warms, it forces some fish to migrate from familiar grounds to unknown new territories and, in the process, local relations also transition. Fishers too change their relationship to place and practices as warming currents bring new species.

Imagining the ocean ecologically through the *intra-actions* of currents with creaturely life, water, lithics and material natureculture, sensitive to the detail and the more expansive planetary view, is 'thinking ecologically' (Code, 2006); and re-thinking relationships 'all the way down'. The temporality of different currents forms relations with particular life forms. For example, at benthic and abyssal depths, the long slow overturning of the thermohaline provides liveability conditions for vent ecologies and delicate nodule life-forms. Following the thermohaline circulation to the ocean's midnight depths is to follow an absent horizon without visibility, where the impacts of material transitions on these life forms is suspended over the long range. Thinking ecologically with the long, slow-time relations of ocean currents signals a need to place time into our considerations (Code, 2006). If relations are foundational to ethics, the transitioning, transcorporeal and intra-active relations of currents and their material solutions signal the need for an ocean ethics attentive to potential material transitions and material memory over difference temporal scales.

Bringing visibility to transition

The slow passage of the thermohaline means that, as warming surface temperatures deepen with our carbon streams, so the risk of long lingering warmth is cast well into the future. The climate change that we affect, and the consequent warming, transition together. Excess carbon emissions impact ocean communities through slow, longer-range and therefore less visible environmental violence that transform liveability conditions. This is the 'image weak' meaning of slow violence described by Nixon (2011), where the slow circulation of deep ocean currents means that consequences of human actions and inputs may not appear for 5, 10 or 100 years or more into the future. Environmental impacts on the remote, deep ocean worlds are also image weak as there is little opportunity to observe changes over time. There is opportunity for feminist new materialist approaches (Alaimo, 2012, 2014; Bennett, 2010; Chandler and Neimanis, 2013; Neimanis, 2017) to sieve the sticky solutions of ocean currents; feel and comb through the slimy, carbon layered sediment; push with the water through open gills, down ridges, across basins, athwart currents and time, to bring into visibility the long-range deep sea relationalities, and create new imaginaries with the potential to bring about positive change.

Things that stir and pump: krill on the move

As discussed throughout this chapter, the ocean is not a static place, fluid or thing that moves from one hold to the next, as if separate from its movement; or merely a habitable vessel for organisms. Its dynamic systems are pluri-dimensional and co-generative with marine organisms and solutions, and with human material expressions and actions, such as carbon excesses and industrial fishing. The agential power of creaturely interactions thrum and vibrate transcorporeally across bodies and waters, as evinced in the collective dynamic power of vertically migrating krill.

Myriad lives emerge daily to squirm, wriggle, pump and glide from seabed to surface, entraining sediment particles as they rise. It is the largest migration of animals on the planet. Searching for food and avoiding predators, they rise in the shelter of night to the moonlit surface: krill, plankton and nekton – threshold-busting creatures ascending from the ocean floor, sea ridges, and through shelf and slope waters. Dawn signals a necessary dive and sinking retreat to the sunless interior. This enormous, daily planetary event expresses the complicated material intra-actions generating oceanic flux and biological mixing (Bianchi et al., 2013; Swadling, 2006). Wilhelmus and Dabiri (2014) propose that vertically migrating zooplankton may input as much energy into the ocean as winds and tides.

Antarctic krill sink once their stomachs are full and rise again to eat a little more, doing this several times during the day (Bianchi et al., 2013; Swadling, 2006). Moving in this way, each krill is an ocean pulse. With several pairs of beating thoracic legs, krill pump up and down the water column daily, creating micro-eddies. Southern Ocean krill numbers are estimated, by weight, at around 420 million tonnes (CCAMLR, 2015). What then of their collective agency in contributing to Southern Ocean movement and biological mixing? How might fisheries management be imagining differently by thinking of the broader material relations of krill and the water column, in all its dimensions. By taking industrial quantities of krill for such things as Omega 3 supplements to ease aching human joints (see Abrahamsson et al., 2015), are we not taking more than the crusty sacs from which to squeeze out a bit of Omega 3? Are we not also removing relational entities and significant contributors to ocean dynamics? Where is an account for loss of ocean movement in the measure of fisheries impact?

Krill fisheries are measured in terms of their seasonally available tonnage, which says little of the dynamic complexity of krill communities through the water column. Their populations are measured only in the upper, more easily accessed layers of the water column but not the deep ocean realms. However, there is evidence of krill dive-bombing deep sea sediment and stirring nutrient plumes up into the water column for circulation (Sanderson, 2008). Imagining additional krill populations at depth suggests the possibility that their collective micro-eddies are an even more powerful biological pump than previously reckoned.

Over centuries of human exploitation of the ocean, how has the loss of marine animals diminished the power of other biological pumps? It is a question made more pressing in the context of climate change and the increasing vulnerability of ocean creatures to transitioning ocean conditions. All levels of the food chain need adequate availability and circulation of nutrients. In earlier centuries when far greater numbers of larger mammals circulated the planet, might another consequence of the loss of marine populations be the collective loss of movement? Few estimates are available of the loss of ocean movement resulting from commercial fishing, though a small number of studies have estimated the energy contributions of whales (Dewar et al., 2006; Lavery et al., 2012). Can we extend the idea of creaturely volition to consider the role of horizontally migrating species in planetary dynamics? Large mammals and fish ride seasonal surface currents from basin to basin. How might their collective muscular outputs contribute to the volition of the currents? I propose that these considerations have potential implications for future fisheries and environmental management. In the context of slowing currents brought on by climate change, how might fisheries management regulate krill fisheries in a way that also accounts for sustainable levels of biological pumping. Further, in an alternative ocean imaginary, might the ocean's dynamic nature be understood through its moving marine creatures?

Antarctic Circumpolar Current projections

The Antarctic Circumpolar Current (ACC) takes us all the way to the icy latitudes of the Southern Ocean, where it moves between Antarctic research vessels, tourist boats, fishing fleets, extreme sailors and whalers. Beneath the ice and chilly dense waters, the ACC sweeps the lively ecologies of Antarctica's broad shelf and Southern Ocean plains and ridges, gliding with communities of Weddell seals, orcas, Emperor Penguins and krill.

Gale forced by the Roaring Forties and Furious Fifties, the ACC circulates uninterrupted around the globe, connecting sea water from the Atlantic, Indian and Pacific Ocean basins. It is an eight-year and 24,000-kilometre journey (Parks and Wildlife Services Tasmania, 2000). The ACC's eastward flow pushes through the extended axis that marks the uneven and, for the most part, unrecognised territorial claims on Antarctica (Rothwell et al., 2015, 740): first Australia, then France, Australia again, New Zealand, then eastward through a large unclaimed area before reaching the territories claimed by Chile, the United Kingdom, Argentina and Norway and then back through Australia. Surface and deep waters flow with a certainty not matched by international conservation agreements. The juridical boundaries of the UNCLOS high seas provisions and Commission for the Conservation of Antarctic Marine Living Resources marine protections are unresolved. Uncertainties associated with conservation of marine living organisms on the Antarctic and Southern Ocean seabed combine with the region's remote and physically difficult conditions to make enforcement of conservation regulations difficult. Unclear conservation regulations generally favour mining companies. For now, though, the steady strength of the ACC and perilous westerlies keep miners at bay, at least as long as technological fixes evade them.

Being barotropic, the ACC reaches all the way to the bottom and, as it moves indiscriminately and relationally across basins, it connects seafloor to sea surface. In a passage describing the ACC's dynamics and sensory responsiveness, Gordon (2001) describes its affective ability to 'feel' the shape of the seafloor. As it feels along the seafloor topography, it projects its journey up into the water column, in equivalent sea surface temperature patterns. How might the watery imprints of other currents, in different locations or times, resonate through the water column? Though not yet near Antarctica's shores, mining machinery, coral colonised drilling pylons, sunken ships, and dumped space craft already scaffold industrialised seafloor-scapes. Thinking with the ACC's affective agency offers another way to imagine an industrial future for the polar seafloor, once favourable mineral prices, advanced technologies, corporate appetite and increasing human populations potentially embrittle the sanctity of Antarctica's environment? How might we speculate on a watery imprint projected by a future ACC as it feels the topography of an industrialised polar seafloor? What new organisms will live in that new imprint and which will be extinguished?

Living with transitioning currents

The transitioning relations of currents and their materiality means that the very definition of what moves changes – outpacing the modelling. Massive eddies off the New South Wales (NSW) coast that were modelled six months ago are entirely different to the ones out there today. Thinking simultaneously with the models and our ocean entanglements draws certain material conditions into present view. No sooner are these perceived, though, than they are already transformed as bundles of transitioning departures and futures carried onward. As François Jullien observes in *Silent Transformations* (2011), neither are transitions all easily discerned, often invisible, unspoken except in the eventual signalling of transformation that they bring about. Transitions are not all imbricated, as if some earlier ocean conditions might return as soon as prevailing conditions are removed. Some transitions signal irrevocable transformation.

Unlike change, which indicates an original condition that might be restored, transitions are continual. Our dependency on a stable climate system makes us vulnerable to the transitional, dynamic agency of currents and their continual change in a direction not yet familiar. Through our technologies, waste and energy consumptions, we are all differently implicated in, and co-generative with, such changes. Through our carbon currents we also alter the pace of change itself. Ocean warming, changing currents, coral bleaching and other marine environmental changes that we see are outcrops of subtle transitions over time, often below the plane of visibility (Jullien, 2011). The extending warm reach of the EAC, the slowing thermohaline circulation and the plastic thickened North Pacific gyre are all just visible signals of material relational qualitative changes.

Closer to shore, currents wash in the ocean's material changes. Those who live with the sea notice fewer pipis burrowing wet sands; fewer shell treasures tangle wrack; stony, tough, mono-toned molluscs and umber-gel algae brown up the rock pools, having out-sustained the vivid coloured anemones and marine plant varieties; and fewer birds scamper the waves. In common with other coastal locations around the world, the ecotones of the New South Wales coastline are still beautiful but very quietly our aesthetic baselines shift with the losses. Carbon thickening briny solutions connect over time with early industrial growth and our persistent material demands.

The 'transitional' in 'change' is a universal processual quality. When we talk of changing currents, changing climates or the changing ocean it is as if the currents, climate and ocean possessed some form of prior integrity, rather than a continual transitionality. What we see or measure as change is the outcrop, the passing signal of transition. Cutting through change to *transition* enables a number of relational connections to be better understood. As a processual quality, transition is shared across natureculture, humans and more-than-human. The materials implicated in transition may also be held or distributed in common, though they may differ in intensity or affect. Transitional currents are an expression of the transitioning ocean, intra-acting with the heat and carbon excesses through which we also transition. Indeed, we transition with, into and through the ocean's materiality. Thinking with the currents in this way necessarily means thinking transitionally.

Transitionality too is responsive to human action. Neimanis invites us to imagine 'watery nature as continuous with human nature, and therefore responsive to human actions' (2014, 10). Connecting our watery natures to ocean currents, we might ask to which human actions we wish them to respond?

The transitional in change

Warming surface waters, storm and rain intensifications are not suddenly upon us, or separable from us, they are our nature/culture transitions, carrying all the way along the intensifying activity of our industry and increasing emissions to the warming ocean. François Jullien (2011) observes that transitions are not easily discerned, often invisible, and unspoken except in the eventual signalling of transformation that they bring about. The slowing thermohaline circulation and increasing presence of plastics are just visible outcrops of subtle transitions already underway.

Transitions are not all imbricated, as if some earlier ocean conditions might return once prevailing conditions are removed. Some transitions signal irrevocable transformation. We talk of changing currents, changing climates or the changing ocean, as if the currents, climate and ocean possessed some form of prior integrity, rather than a continual transition. Cutting through change to transition enables certain relational connections between bodies to be more closely understood. As a processual quality, for example, transition is shared across nature/culture, humans and more-than-human.

How might we imagine the material conditions for ocean transitions differently, with a view to the manifold liveliness of the world? How might we develop ecological thinking and governance approaches that are more sensitive and adaptive to shared material vulnerabilities, over time? Sarah Ensor identified in Rachel Carson's work an ecocritical approach that sees the present as 'the future of any number of pasts – some near, some far, some recent, some long gone' (Ensor, 2012, 418). It is in essence an engagement with transition. Repurposing Carson's approach for ocean governance, we might consider the ocean's present condition and our material additions and extractions as already legible futures. If the conditions experienced by today's ocean are already the pasts of any number of futures, all the more reason for intensified precaution and longer-range temporal scales to inform our governance regimes.

The ocean's transitions are silent transformations that are not easily seen and for which there are no utterances yet shaped to respond. The ocean's silent transformations may be either too far away, unfolding over the long range of the thermohaline, or more contemporaneously, through the extending warm streams of the EAC; or the gradual removal of ocean pulses too small to see. Missing the silent material transitions of our material inputs and extractions we are, as Jullien (2011) observes, surprised or bewildered when the visible effects appear.

Notes

1 Broadly conceived, these encompass such forms as jet stream, gyres, eddies, cast-off rings, overturning, upwellings and myriad biological pumps.
2 This chapter uses the word ocean to capture the physical interconnectedness of lively oceanic worlds.

References

Abrahamsson, C. S., Bertoni, F., Mol, A. and Ibáñez Martín, R. (2015). Living with Omega 3: New materialism and enduring concerns. *Environment and Planning D: Society and Space*, 33(1), 4–19.

Alaimo, S. (2010). *Bodily Natures: Science, Environment and the Material Self.* Indiana, USA: Indiana University Press.

Alaimo, S. (2012). States of suspension: Trans-corporeality at sea. *Interdisciplinary Studies in Literature and Environment*, 19(3), 476–493.

Alaimo, S. (2014). Oceanic origins plastic activism and new materialism at sea. In S. Iovino and S. Oppermann (Eds.), *Material Ecocriticism*. Indiana, USA: Indiana University Press. pp. 186–203.

Barad, K. (2007). *Meeting the Universe Halfway: Quantum Physics and the Entanglements of Matter and Meaning*. Durham and London: Duke University Press.

Bennett, J. (2010). *Vibrant Matter: A Political Ecology of Things*. USA: Duke University Press.

Bianchi, D., Galbraith, E. D., Carozza, D. A., Mislan, K. A. S. and Stock, C. A. (2013). Intensification of open-ocean oxygen depletion by vertically migrating animals. *Nature Geoscience*, 6(7), 545–548.

Bosselmann, K. (2010). Losing the forest for the trees: Environmental reductionism in the law. *Sustainability*, 2(8), 2424–2448.

Brierley, A. S. and Kingsford, M. J. (2009). Impacts of climate change on marine organisms and ecosystems. *Current Biology*, 19(14), 602–614.

CCAMLR. (2015). Krill – biology, ecology and fishing | CCAMLR. [Online] https://www.ccamlr.org/en/fisheries/krill-%E2%80%93-biology-ecology-and-fishing Accessed 3 July 2017).

Castoriadis, C. (1987). *The Imaginary Institution of Society*. Cambridge: Polity Press.

Chandler, M. and Neimanis, A. (2013). Water and gestationality: What flows beneath ethics'. In C. Chen, J. Macleod and A. Neimanis (Eds.), *Thinking with Water*. McGill: Queens University Press. pp. 61–83.

Code, L. (2006). *Ecological Thinking; The Politics of Epistemic Location*. New York: Oxford University Press.

Code, L. (2012). Ecological responsibilities: Which trees? Where? Why? *Journal of Human Rights and the Environment*, 3, 84–99.

Cressey, D. (2016). The plastic ocean. *Nature*, 536(7616), 263–265.

De Lavergne, C., Palter, J. B., Galbraith, E. D., Bernardello, R. and Marinov, I. (2014). Cessation of deep convection in the open Southern Ocean under anthropogenic climate change. *Nature Climate Change*, 4(4), 278–282.

Dewar, W. K., Bingham, R. J., Iverson, R. L., Nowacek, D. P., St. Laurent, L. C. and Wiebe, P. H. (2006). Does the marine biosphere mix the ocean? *Journal of Marine Research*, 64 (4), 541–561.

Earle, S. (2016). Deep sea mining: An invisible land grab – Mission blue. Mission Blue. [Online] https://www.mission-blue.org/2016/07/deep-sea-mining-an-invisible-land-grab/ (Accessed 5 December 2016).

Eissa, A. E. and Zaki, M. M. (2011). The impact of global climatic changes on the aquatic environment. *Procedia Environmental Sciences*, 4, 251–259.

Ensor, S. (2012). Spinster ecology: Rachel Carson, Sarah Orne Jewett, and non-productive futurity. *American Literature*, 84(2), 409–435.

Gordon, A. L. (2001) Current systems in the Southern Ocean. In *Encyclopedia of Ocean Sciences*. Oxford: Academic Press, pp. 613–621. [Online] https://www.sciencedirect.com/science/article/pii/B012227430X00369X (Accessed 30 May 2018)

Grear, A. (2015). Towards new legal foundations? In search of renewing foundations. In A. Grear and E. Grant (Eds.), *Thought, Law, Rights and Action in the Age of Environmental Crisis*. Cheltenham: Edward Elgar Publishing. pp. 283–313.

Haraway, D. (2003). The Companion Species Manifesto: Dogs, People and Significant Otherness. Chicago: Prickly Paradigm Press.

Helmreich, S. (2009). *Alien Ocean: Anthropological Voyages in Microbial Seas*. Berkeley, CA: University of California Press.

Helmreich, S. (2014). Waves: An anthropology of scientific things (The 2014 Lewis Henry Morgan Lecture). *HAU: Journal of Ethnographic Theory*, 4(3), 265–284.

Hill, S. (2013). A view to a krill: Warming seas may leave predators hungry. [Online] http://theconversation.com/a-view-to-a-krill-warming-seas-may-leave-predators-hungry-17383 (Accessed 7 August 2015).

Hunter, T. and Taylor, M. (2013). *Deep Sea Bed Mining in the South Pacific (Background Paper)*. Centre for International Minerals and Energy Law: Brisbane: University of Queensland.

International Seabed Authority. (2011). *Environmental management of deep-sea chemosynthetic ecosystems: Justification of and considerations for a spatially-based approach*. ((ISA Technical Study Series No. 9). Kingston, Jamaica: International Seabed Authority.

Jullien, F. (2011). *The Silent Transformations*. London: Seagull Books.

Keller, K., Tan, K., Morel, F. M. M. and Bradford, D. F. (2000). Preserving the ocean circulation: Implications for climate policy. *Climatic Change*, 47(1–2), 17–43.

Kelly, K. (2011). *The Inertia Trap*. Canberra: Ronin Films.

Lavery, T. J., Roudnew, B., Seuront, L., Mitchell, J. G. and Middleton, J. (2012). Can whales mix the ocean?. *Biogeosciences Discussions*, 9(7), 8387–8403.

Mann, K. H. and Lazier, J. R. N. (1996). *Dynamics of Marine Ecosystems: Biological-Physical Interactions in the Oceans* ((Second edition)). Oxford: Blackwell.

Milman, O. (2015). Australian fish moving south as climate changes, say researchers. *The Guardian*. [Online] http://www.theguardian.com/environment/2015/jan/29/australian-fish-moving-south-as-climate-changes-say-researchers (Accessed 23 February 2016).

Mitchell, A. (2009). *Seasick: The Hidden Ecological Crisis of the Global Ocean*. Chicago: University of Chicago Press.

Neimanis, A. (2013). Feminist subjectivity, watered. *Feminist Review*, 103, 23–41.

Neimanis, A. (2014). Alongside the right to water, a posthumanist feminist imaginary. *Journal of Human Rights and the Environment*, 5(1), 5–24.

Neimanis, A. (2017). *Bodies of Water: Posthuman Feminist Phenomenology*. London: Bloomsbury.

Neimanis, A., Åsberg, C. and Hayes, S. (2015). Post-humanist imaginaries. In K. Bäckstrand and E. Lövbrand (Eds.), *Research Handbook on Climate Governance*. Cheltenham: Edward Elgar Publishing. pp. 480–490.

Nixon, R. (2011). *Slow Violence and the Environmentalism of the Poor*. Cambridge, MA: Harvard University Press.

Parks and Wildlife Services Tasmania. (2000). *Antarctic Circumpolar Current*. Hobart, Tasmania: Parks and Wildlife Services.

Passow, U. (2015). What happened to the oil from the deepwater horizon spill? 'Marine snow' provides a clue. *The Conversation*. [Online] http://www.iflscience.com/environment/what-happened-oil-deepwater-horizon-spill-marine-snow-provides-clue/(Accessed 10 June 2015).

Pauly, D. (2010). *5 Easy Pieces: The Impact of Fisheries on Marine Ecosystems*. Washington: Island Press.

Plumwood, V. (2002) *Environmental Culture: The Ecological Crisis of Reason*, London: Routledge.

Probyn, E. (2016). *Eating the Ocean*. United Kingdom: Duke University Press.

Rosenbaum, H. and Grey, F. (2015). *Deep Sea Mining: Out of Our Depth: A Critique of Nautilus Minerals Report on the Solwara 1 Project*. Deep Sea Mining Campaign. [Online] http://www.deepseaminingoutofourdepth.org/a-critique-of-nautilus-minerals-report-in-solwara-1-project/(Accessed 20 December 2016).

Rothwell, D., Elferink, O., Scott, K. and Stephens, T. (Eds). (2015). *The Oxford Handbook of The Law of The Sea*. Oxford: Oxford University Press.

Sanderson, K. (2008). Krill spotted diving at depth. *Nature*. [Online], https://doi.org/10.1038/news.2008.620. Accessed 28 February 2015)..

Scher, H. D., Whittaker, J. M., Williams, S. E., Latimer, J. C., Kordesch, W. E. C. and Delaney, M. L. (2015). Onset of Antarctic Circumpolar Current 30 million years ago as Tasmanian Gateway aligned with westerlies. *Nature*, 523(7562), 580–583.

Steinberg, P. (2001). *The Social Construction of the Ocean*. Cambridge: Cambridge University Press.

Steinberg, P. and Peters, K. (2015). Wet ontologies, fluid spaces: Giving depth to volume through oceanic thinking. *Environment and Planning D: Society and Space*, 33(2), 247–264.

Swadling, K. M. (2006). Krill migration: Up and down all night. *Current Biology*, 16(5), 173–175.

UNEP. (2014). *Sand: Rarer than one thinks* (UNEP Global Environmental Alert Service). United Nations Environment Programme. [Online] http://na.unep.net/geas/archive/pdfs/GEAS_Mar2014_Sand_Mining.pdf (Accessed 5 August 2017).

Van Gennip, S. J., Popova, E. E., Yool, A., Pecl, G. T., Hobday, A. J. and Sorte, C. J. B. (2017). Going with the flow: The role of ocean circulation in global marine ecosystems under a changing climate. *Global Change Biology*, 23(7), 2602–2617.

Van Sebille, E., Oliver, E. and Brown, J. (2014). Can you surf the East Australian Current, Finding Nemo-style? [Online] http://theconversation.com/can-you-surf-the-east-australian-current-finding-nemo-style-27392 (Accessed 22 December 2015).

Wilhelmus, M. M. and Dabiri, J. O. (2014). Observations of large-scale fluid transport by laser-guided plankton aggregations. *Physics of Fluids*, 26(10), 101302.

Worm, B., Barbier, E. B., Beaumont, N., Duffy, J. E., Folke, C., Halpern, B. S., . . . Watson, R. (2006). Impacts of biodiversity loss on ocean ecosystem services. *Science*, 314(5800), 787–790.

Part Two

Engagements and experiences

9   Seafarers and work

Endless, sleepless, floating journeys: the sea as workplace

Maria Borovnik

I like to go to sea.
What is it that you like about going to sea?
I enjoy the sea when it's going like this [he moves his hands and body to show].
Oh. You like the ship when it's a bit wobbly, moving to this side or that side?
And there are also many different kinds of oceans.
You find the oceans look different from each other?
Yes. You can see the difference. Like – [in] some places, the ocean looks like green. Not like blue. Or, it's not clear. Some places, like Shanghai! It looks like mud.
Really?!
Yeah. It's not green. It's not blue. It looks like mud. So, I see the difference from our ocean [the Pacific]. It's much different.
The ocean in the Islands is different. It is crystal clear.
Also deep. Very deep.
[Silence]

(Interview with Seafarer Z)

For a moment there was something transcendent about sitting together and recalling the beauty of the sea, of the different oceans. Not many seafarers shared such deep feelings and the sea itself was fairly absent in the imaginations of most seafarers I had spoken with. Searching my interview material, comprising 15 years of research, for anything seafarers might have said about the beauty, the colour or the shape of the sea while on ships brought up very little. The material, however, revealed that seafarers might be *feeling* the sea through the medium of the ship in which they work and live rather than just *seeing* it or *hearing* it. Therefore, this chapter focuses upon the sensual and corporeal experiences of seafarers. These experiences are not separated from the environment. This chapter will explain that rather than being at a distance, seafarers are a part of the ocean, and are 'emplaced' within it (Ingold, 2011). The seascape for those sailing from port to port is not 'tied to any specific sensory register' (Ingold, 2011, 136) while focusing on keeping the ship on an even keel and on track, protecting the decks from the salty water, and ensuring the engines are running smoothly. It is rather a scape 'beyond the visual', including both the human and non-human world (Brown, 2015, 20). More exactly, the sea is experienced as engaging in a 'meshwork' of senses, elements and materials (Ingold, 2004). Such interactions with the seascape are shaped by socio-cultural understandings that are brought into this floating workplace and are constantly constructed or reconstructed by day and night activities and tasks. This chapter explores how seascapes, through which seafarers move and in which they engage, are constituted by imaginaries, memories and, in turn, how seafarers contribute to the sea as a place with their own material, sensual and imaginary actions.

The material presented in this chapter draws on interviews mainly with seafarers from Kiribati and Tuvalu. Some data was also taken from interviews with a small number of seafarers from the Philippines, Ukraine and New Zealand that I met during this voyage. This data was collected from a long-term research project in the Pacific and during a 29-day voyage with a crew of 21 on a containership in December 2009 and January 2010. I developed good relationships spending time with seafarers during their activities on deck, in the engine room, in the mess or galley area, or on the bridge, and during after-hour happenings. For reasons of confidentiality I have taken care not to reveal the exact nationality and have also generalised (or left out) department and job types where possible.

This chapter deals with multiple perspectives of the sea as a work environment, focusing particularly on the material and emotional interactions between the watery ocean environment, the accompanying sea and landscapes, and people dealing with the flow of *within* and *in-between* experiences. The following section will conceptualise these multifaceted, syntonic encounters between seafarers, the ship's materialities and the unpredictable, fluid elements of the seascape.

Ship/life/sea

On their journeys from port to port, seafarers not only encounter, but are deeply encapsulated in their ocean environment in such a way that the meshwork of materiality and senses resonate with each other (Steinberg and Peters, 2015, 250; Ingold, 2004). For seafarers in particular, being immersed in ongoing loops of merchant activity is a 'continuous and never-ending process' of weaving life with seascapes (Ingold, 2004, 333; see also Cook and Edensor, 2017), in which seascapes are mobilised, articulated or performed. Seafarers are both actively and passively involved in these performances and the 'affective atmospheres' of ship journeys (Ashmore, 2013, 596; Anderson, 2009). They are core actors in making trade possible and at the same time are entrapped in its global economies (Borovnik, 2011a; Martin, 2013; Parker, 2013; Squire, 2015). Seafarers are, however, mostly on the periphery when it comes to labour mobility, taking up jobs that involve a certain type of hardship (Birtchnell and Urry, 2015; Urry, 2014), and insecurity (Langewiesche, 2005).

Many narratives by seafarers tackle the complexities of the economic necessities of providing for families at home (see Borovnik, 2011b; Fajardo, 2011; Sampson, 2013). Therefore, in the merchant seafarers' context the sea is mainly symbolised as a 'grounded centre of capital' (Anim-Addo et al., 2015, 341, drawing on Steinberg, 2001), a seemingly 'friction-free' space facilitating transport of goods and people – at high speed, in a globally competitive manner (Steinberg, 2001; Martin, 2013). In this economic context, seafarers, while facilitating transport of consumer goods, usually remain invisible (George, 2013; Urry, 2014). Joining these two different perspectives of sensuality and economic necessity, two notions of mobility converge in the lives of seafarers. One notion is that of *transport* through the ships that are navigating consumer goods from port to port, dealing with a schedule determined by cargo agencies, taking goods from sites of production to sites of consumption. Ships in this perspective are nodes of a complex assemblage. The sea from this perspective is also a node, a temperamental facilitator of capitalism, making it happen or slowing it down. Seafarers are inevitably intertwined with the rationale of transport that is driven by the immanence of capital and trade. Signing up for crewing jobs, designed to maximise profitability for shipping companies that are, in turn, providing for the livelihoods of families at home, means also signing up to immobility and temporary lack of freedom of movement as part of the deal in this transport environment.

Another notion to consider is *wayfaring*. Seafarers as wayfarers are experiencing, discovering and knowing through movement as temporary and unfolding. Wayfaring is a 'coming-into-sight and passing-out-of-sight' of surfaces rather than a completed sequence of settled movements from one fixed point to another (Ingold, 2000, 239). In addition, one could argue that movement on ships also includes coming-into-*sound* and passing-out-of-*sound* and a continuation of sequences of kinaesthetic (embodied, felt or sensed) sensations. Moving across the sea requires seafarers to respond to the 'movements of the ocean or sea that are determined by forces external to the water: the wind, the temperature, the gravitational pull of the sun and moon, the solid seabed and its depth' (Peters, 2015, 266). Over time, by focusing on daily work-related tasks that facilitate transport, seafarers acquire and deepen their knowledge about the oceans, their rhythms and unpredictability – or 'tune in' to the environment, as Hunt (2018) would put it. The lengthy and repetitive voyages move seafarers through a space in which time becomes fluid, and the journey itself is of secondary importance; rather, the everyday focus is on what is done, what needs to be done, and how things can be done on a moving ship. As a consequence, knowledge unfolds in the processes of being, sensing and doing, of floating, steering and mooring. From this perspective the interaction between sea/life/ship/shore is a roundel (rather than a point, or blob as it would be from a transport perspective) (Ingold, 2015, 7), a very complicated accumulation of lines in which the different materialities, socialities and sensualities keep moving. Wayfaring and the acquisition of knowledge for seafarers in this process is a rhizomic formation, one that connects and links qualities of different natures without any actual beginnings and ends (Deleuze and Guattari, 1987, 21). Applying this perspective that emphasises the 'continuous formation' of spaces as we move through them opens the opportunity to view the environment as being physically and socially intertwined with those that are in it – including the ship (Ingold, 2000, 230). The ship, then, is a space that is 'part of a wider global fabric or *meshwork* of movement; of ties and knots forging places, times and experiences' (Anim-Addo et al., 2015, 341, drawing on Ingold, 2011). The sea in this view is part of this fabric.

In acknowledging seafarers as economic and emotional parts of seascapes brings up the possibility of exploring how these two different frameworks are intertwined. If seafarers' experiences could be deconstructed into functional, economic components only, they then would fit the assemblage framework well. However, bringing in sensual and corporeal dimensions makes the understanding of ship/life/sea more complex. Bissell (2010) suggests that bodies are more than purposeful, functioning producers of activity and that embodied experiences of movement and the material initiating mobility opens up new ways of thinking. Drawing attention to the everyday life interactions and including the multi-sensorality and corporeality involved can assist in unfolding some of the complexities of mobile lived experience (Pink, 2015; Fajardo, 2002). The sensual and embodied is a focus in Fajordo's (2002) study on Filipino masculinities. Both Fajardo and Rigg (2007, 16) underline the importance of the everyday and non-representational in studies on globalisation. By paying attention to 'ordinary people' and 'everyday action', value is added. This focus is particularly important considering the overlapping events of globalisation and development, which would usually normalise aspects of transport. Yet wayfaring is an experience of the everyday, where seafarers are experienced and sensual actors in the process.

Endless journeys

Not feeling the sea is often more important, more bearable, than being with it. When seafarers sign up for their ship work, they actively agree to work in a constrained work environment with an unknown future, exposed to a possible 'watery grave'. As Urry describes it:

> Up to the twentieth century all international transport involved moving objects and people upon the surface of these dark and dangerous oceans. Those venturing to sea, the seafarers, often came to grief, as dangerous storms, mysterious currents, underwater rocks and marauding pirates drove many travellers and their goods to a 'watery grave'. For pretty well all of human history there has been an 'outlaw sea' that engendered danger, disease and death.
>
> (Urry, 2014, 157).

Seafarers resign themselves to an environment and future that depends on the temperamental conditions of wind and currents, unknown situations, and on the decisions made by shore-based companies. Urry argues that the seas are in process, but they are also 'unruly because of what is happening onshore' (2014, 158). The shipping transport industry, with the introduction of flags of convenience, has provided opportunities and also enhanced global competition among seafarers. Working on ships that are flying under a foreign flag or the second register of a European nation state offers flexibility for shipping owners with regard to negotiating wages and the provision of social security aspects of contracts. As Langewiesche (2005, 4) notes, 'ships themselves are expressions' of a system that is bound by profit making. Ships move about independently, 'many of them without allegiances of any kind, frequently changing their identity and assuming whatever nationality – or "flags" – allows them to proceed as they please' (Langewiesche, 2005, 4). Depending on the nationality, union agreements and manning agency, conditions will vary and it is possible that crew members of different nationalities working at the same level might work different lengths of contract and have divergent salaries (Alderton et al., 2004; Borovnik, 2011a, 2011b).

Under current global dynamics, seafarers consequently work in the in-between of shore-based decision-making and the unruly behaviour of ship and sea. During the last couple of decades, their job has become increasingly stressful: first with the demand to increase speed; then with the introduction of surveillance and patrol systems both on shore and on ship. Stress increases with each economic crisis where shipping owners are forced to reduce crew numbers. Seafarers can feel immobilised without opportunities to go ashore, and lack of sleep as a consequence of the additional work to make up for crewing shortages (Borovnik, 2011a; Sampson, 2013). Stress and burnout syndromes were recognised, in an explorative literature review by Mellbye and Carter (2017), as the strongest factors for suicide, and numbers seem to be rising. There are two types of stresses that seafarers face. One is caused by frequent turnarounds, which requires longer watch-keeping hours; at the same time, alertness is needed during port operations, to get the ship ready to go again. This kind of stress is higher when turnaround times are short and when there are few opportunities to relax ashore (George, 2013; Langewiesche, 2005; Sampson, 2013). A second, quite different type of stress observed by ethnographers, however, is social isolation (Thomas, 2003), usually experienced more during longer voyages, when there is 'nothing but ocean all around, a very pretty sight but … these are the loneliest times, when there is nothing but sea' (Journal entry, 3 January 2009; in Borovnik, 2011a, 67). Land- and seascape are interwoven in the meshwork of economic and sensual realities of monotony, excitement and restlessness – where seafarers are not just working but also 'being jiggled about' over long periods of time (Bissell, 2010, 480). Conscious or subconscious awareness of uncertainty, loneliness, stuckness and tiredness make journeys seem endless. Not feeling anything – numbness or distractions – are key strategies employed when facing this monotony that is part of the realities of seafaring.

For many younger seafarers, going to sea is attractive as this activity offers the opportunity to gain experiences beyond their local environments and to see the world. Seeing the world means exploring different shores, and to some extent seeing the shore from the perspective of the sea. Ingold suggests that by taking the perspective of 'sea-ing the land ... it is the solidity of the ground itself that is thrown into doubt. That is also restless, in ceaseless motion and change' (2011, 131). After a long time in the job, seafarers frequently sense they have nothing to talk about with land-based family and friends. Experienced seafarers also feel as though they are fairly detached from land-based events at home and worldwide. Even though in recent years some seafarers may have access to computers and daily news summaries aboard, many might not have this privilege or are too busy, or tired, to follow updates. Experienced seafarers might feel resigned to the repetitive and harsh reality of their work on board ship, but being drawn back to the sea seems inevitable for many (Borovnik, 2011a). Being on the ship – a space that has been described as an unreal, transgressive, hyperspace (Sampson, 2003; Stanley, 2008), or fluid, marginal heterospace (Foucault, 1994) – involves seafarers sensing a detached, unreal, marginal, and yet familiar, reality. However, when sea-ing *land*scape from the perspective of ship based reality, it also becomes unreal, as several of the experienced seafarers told me during my journey.

> One of the engineers says that even after one month of the ship I will notice that my life will have changed, that I will have a different way of looking at things on shore. He cannot explain it. After some digging, he says that I will be running around with a big smile on my face because I will be so happy to be on shore. He says that life on ship is so very different from everything else he knows. Hard to explain, but it is out of this world, such a different world in itself. After seven months, he says, or imagine after 11 or 12 months you are a different person. Things will then always be different for you ... Later when I visit the engine room I contemplate his words. The engine room does not have windows. There is no sign of the outside world whatsoever, other than the option to send an email. The engine room also moves, and today there was quite a lot of weather outside, rain and rough sea – actually very beautiful. And so the engine room goes up and down quite a lot.
>
> (Journal entry, 11 January 2010)

Experiencing the ship for a length of time is a key trigger experience affecting the emotion of happiness that this engineer referred to. Carrying the memory of the sea and the onboard experience will change the perception of shore life and 'things will be always different' or will be looked at differently. These differences are complex as they do involve the moving, the sensing of the sea, as well as the 'sea-ing'. In the following example, happy feelings and excitement are expressed – these extending beyond simply having a break and the opportunity for shore leave. Actually being on the land itself after long periods on sea makes this young seafarer feel good. In this example, land (and not the sea) becomes a symbol of freedom.

Do you go on shore sometimes?
That's the one! Yeah. That's the one that makes you happy, eh?! You enjoy!
  You enjoy! After leaving the ship some of the feelings you feel!

You enjoy going on shore.
[Laughter]
What kind of shore do you. . .?
Any! Any is good . . . But the feeling! When you go ashore! You feel good.
  You enjoy.
What makes you feel so good when you go ashore?
Because [you're] always on the ship. At sea. Always, always. But that . . . the
  first time you go out . . . [Whispers] The feeling. It's a good feeling.

(Interview with Seafarer H)

Having only spent a few weeks on 'my' containership – and not a few months –
did place me as a land-based or land-biased person, one who was only slowly
transitioning to live at sea. My experience traversing the Indian and South China
Seas felt fairly fluid once time zones had to be crossed and clocks needed adjusting
several times, and the ship moved through extreme climates within only one or two
weeks. I also felt that day- and night-times were not as clear-cut as during my usual
shore life. The vessel could reach a port destination anytime, and waiting for a
pilot often lasted several hours and brought turnaround operations into night-time.
These sensations of timelessness were pleasant to me, but I was being made aware
by some men that this pleasant excitement of newness would change over time:

> One of the officers explained to me that if I would stay on this ship for more
> than four months even I would feel bored. For him, at the beginning it was
> all quite exciting, like it is for a little kid. Everything was new, and colourful
> and interesting. Time seemed to stretch for too long, summer seemed
> endless. But now things are different for him. Days are all the same. Work
> is all the same . . . And there is not much time for sleep.
>
> (Journal entry, 7 January 2010)

Shift work and the resulting day- and night-time fluidities affect sleep hours.
Those that get time off when in port will have to make a choice between sleep
and enjoyment – often sleep will be more important, as spending time ashore will
take a toll on tiredness. Those that are on watch-keeping duty will be at this task
longer than usual during port times. Landing during night-time requires additional alertness and readiness to help with cargo loading and unloading, or with
any repair-related jobs in the engine department (Borovnik, 2017). Also, the
ship's tremulous materiality and sound environment makes sleep difficult. During
port times, cranes and shifting containers keep cracking and banging. People
running up and down stairways, chatting, shouting or laughing, add to the
soundscape, or *ensound* people, as Ingold (2011, 139) would put it. During
seagoing times, sounds are caused by the changing sea behaviours: furniture
squeaks; there will be rattles from metal against metal outside somewhere; and
machines are making a soothing, but ongoing humming noise. The movement of
the ship is unpredictable and even during smooth weather, lovely rocking
sensations are interrupted by irregular commotions, bangings or crashes. Bissell

(2010, 479) noticed that kinaesthetic affects, such as vibrations 'seem to permeate the embodied experience' and become part of 'being thrust through a landscape'. Such vibrating (jiggling) sensations, emanating from the actual movement of the ship can be 'mildly' uncomfortable or irritating and eventually become unsettling and tiring (Bissell, 2009). Fatigue is probably the most shared and common feeling on ships. Being tired is, under the circumstances, better than being agitated. The hierarchical system on ships – the need to follow orders for the sake of ship safety and good performance reports that could affect re-contracting or promotion prospects for those that work on lower scales of the hierarchy – causes some degree of resistance, resentment or resignation. Sleepy, exhausted bodies affect the way a seafarer feels and senses the environment. Lack of sleep makes everything dreary and dull. Reactions (and possible emotionally driven over-reactions) may slow response times and people can become more accident-prone.

Not feeling the sea, the numbing of one's feelings, awareness and reactions, are then facilitators, enabling seafarers to face difficult realities. These may be conscious, or possibly more subconscious strategies of many of those working on ships. Numbing helps to avoid thinking about helplessness or powerlessness, and the continuous, underlying awareness of a possible 'watery grave' (Urry, 2014). For example, one of the seafarers I met on 'my' ship explained to me:

> It is impossible for him to go asleep with all the concerns going on in his head of what could happen and maybe they could happen and so on … He therefore tries to kill some time by staying awake even off-duty and hanging out with others.
>
> (Journal entry, 25 December 2009)

Not to feel the sea and its uncertainty, and not to overthink any work-related problems, requires distraction. This includes staying awake with the consequence of feeling over-tired adding to deadening (numbing) worrying feelings. Other ways of facing the reality are distractions through mingling with others, or praying. Uncertainties in seafaring jobs are within the intersection of economic concerns and concerns for life. Many of those I talked to had a similar view to this young man on how it is to be a seafarer:

> The best thing about being a seaman is the salary. The worst thing for me is, you go travel, you know, very far, far away and you don't know what will happen to you … You sign up for a dangerous job … It's like [being] a soldier. You go to a war. You don't know: can I come back to my family? You always pray, always pray.
>
> (Interview with Seafarer P)

This quote reflects on the global reality of most seafarers. Taking up such a job brings great opportunities for home and family – ultimately opportunities for their children, whom many seafarers hope will not have to work in this unpredictable environment. The comparison to the life of a soldier demonstrates this commitment to home and family. It also is a synonym for the power hierarchies experienced on ships, and the ultimate work–life framing power of the ocean. Life becomes unpredictable and one can be hit at any moment by unexpected danger, weather, pirates, fire, ropes or falls. An even higher power must be trusted. Another seafarer I talked to compared his job to astronauts going into the unknown of space;

> He said that when astronauts are prepared going to space they are prepared, they have so many requirements and standards ... but when a seafarer signs up for a contract to go on board a ship, he never knows what is going to happen next. Will there be dangerous weather, like typhoons, or will there be something wrong with the ship, or other dangers. Will they get on with the crew and what kind of captain will it be and so on. He says, it is totally crazy.
>
> (Journal entry, 4 January 2010)

In order to cope with these inner feelings of underlying danger, seafarers might choose to focus strongly on the work at hand every day, and force themselves not to *feel* the feelings. Focusing on a continuum of work and distraction does take away from looking at the sea too closely. Trust, however, is the opposite coping strategy. While numbing is deadening one's feelings, trust is life-assuring. Trust may be focusing on a higher power which can be reached through prayers. Yet trust can also involve an inner belief in technology and the expertise of senior officers in knowing how to handle this technology – to be competently controlling nature and bringing order into chaos. Stormy weather and the possible consequences may bring life and excitement back into the monotony of the journey. The following quote is a continuation of my interview with Seafarer Z, who quite delightfully pointed out that rough seas are welcome in his views as long as the ship is built and maintained by modern technology and steered by experts:

You enjoy the water.
Because in [the Islands] we never travel on the ship in very bad weather. We never travel in bad weather because the ships cannot sustain during bad weather. But here, overseas, never mind. We can still travel. So I can see now, enjoy the ... – And every time when there is a rough sea I wake up, because in my cabin some things go down. I go down to the galley and I see everything down. It's going like this [he laughs; his face is beaming and he waves his hands].
It is going backwards and forwards.
And I ... it is very dangerous. Because this is my smallest ship. Because most of my ships were very big ships. And they can also go like this [again waves his hands].
Big ships can also go in very tough ...
Yes. Like this. And very deep. And our ships in [the Islands] are very small. Maybe [he laughs] they are already capsized (in this kind of weather).

And you think a ship like this one can capsize as well?
I think it can, but they will never because they always go like this [shows] and
always [come back]. But I think they say it only fails when it fails stability.
Then it can capsize. But you know the Chief Mate and Captain are very
professional.

(Interview with Seafarer Z)

Recalling good memories, such as the interesting colours of the different seas he
has travelled to, reminds Seafarer Z of home, which then triggers memories of
excitement, then fear and then reassurance. To a certain degree he feels excitement when something changes, that interrupts the monotony of daily work, but
then he compares rough sea situations with his home experiences and realises his
vulnerable position aboard a ship. He then reassures himself that considering the
advanced technology and professional expertise of ship senior officers will
counteract against any possible life threatening danger. The most important
issue to prevent dangerous situations is the stability of a ship and a need for
balance.

Floating and balance

The ship sometimes changes very suddenly.
Yeah. And that time, a very big swell [comes in] – you cannot walk.
*And even when the sea is calmer, the ship can go like this [makes a bending
movement with her body]. Very suddenly sometimes. I noticed that when I
walk upstairs and carry a cup of tea that it can suddenly go like ...*
Yeah, yeah. And spill the tea with you. That's our life.

(Interview with Seafarer P)

The sea is powerful and technology has attempted to control this awesome,
magnificent wilderness and tame it into *emporion* – a bridging, enabling environment (Borovnik, 2005). And yet this grand nature will continue to battle with
transport technology. Containerships are fairly tough, huge in size, moving
technologies – but they are also vulnerable and need ongoing tuning and adjust-
ment to keep homeostasis, to stay balanced. People also must develop sea-legs to
stay in balance. Adjusting ship technology to aid the ship's mobility must be done
on all major levels. Firstly, in port, when containers are being loaded or unloaded,
their different weights have to be adjusted and the ship's balance is electronically
checked on an ongoing basis. Balance regulations are achieved by pouring water
into a large piping system on each side to counterbalance containers that are being
taken off, and letting out water on opposite sides of where containers are being
placed. Secondly, engines must be kept in mint condition and monitoring of usage
of each part is needed to keep the ship running smoothly. Thirdly, the ship
movement itself needs constant fine-tuning through steering to stay on course.
This is done by continuously adjusting and countering any drift currents (Peters,

2015). Finally, a combination of map-reading and computer monitoring of the sea is needed, to discover any possible obstruction well in advance (Hutchins, 1996). Neglecting any of these balancing components can cause disruption if not danger, something the following young seaman explains:

> The compass is moving if the ship moves. If the ship goes straight ahead, things are OK, but if you move just a little bit [laughter], it's like arrhrr-rhuoah ... capsize! So you try make it [the steering wheel] steady, not try to turn too hard. Just a bit slowly, ooohmmm, steady.
>
> (Interview with Seafarer H)

Here I noted the use of sounds, such as 'arrhrrrhuoah' to express imbalance and 'ooohmmm' to express the smooth and steady steering and movement. These sounds are an articulation of the actual sensation that this man experiences on board his ship and how the ship responds to the constant flux and processes of the water it deals with (Vannini and Taggart, 2014). This man remembers these sensations not just by *visualising* the movement, but by *soundalising* it too. This sound memory is very closely linked to the memory of the kinaesthesia of movement. In fact, the kinaesthetic effect of the ship's movement, through subtle or sudden steering is experienced, memorised and articulated as sound. For Ingold, the environment that is perceived by any of our senses is not split into different kinds of worlds or experiences:

> [T]he environment that we experience, know and move around is not sliced up along the lines of the sensory pathways by which we enter into it. The world we perceive is the *same* world, whatever path we take, and in perceiving it, each of us acts as in undivided centre of movement and awareness.
>
> (Ingold, 2011, 136)

The embodied registers mostly used by seafarers in response to ship behaviour are perceptions of kinesthesia, touch, sound and smell. Drawing on Attlee (2011, 5), Edensor (2015, 561) acknowledges the practical competence of navigators who use subtle senses in their practice, including sound, smell and tactile senses, particularly during night travels. These senses operate in, as Martin has put it, an 'entanglement of elemental and bodily forces' that 'fold through each other' (2011, 455, drawing on Whatmore, 2002, 119), especially in foggy or stormy weather. In the section above I explained how the underlying feelings of being enclosed by ship and sea, the helplessness and the possible dangers this boundedness entails, are dealt with by mostly numbing oneself. Once some of these dangers become reality, a combination of different sensations is being triggered. The following examples will illustrate how. Earlier I noted how the beginning of danger might bring up some initial excitement that will soon be followed by realities of vulnerability. Typhoons or storms are extreme dangers to ship and life. Seafarer H remembered a time when his ship was hit by a storm as being very scary:

That was a very bad time. The ship [was] going like this and [was] pitching
    like this: pshshsh pashewww boow wooh! Ooh, arh! You can't sleep. No
    sleep. Everything is going bahh bahh bahh bahh bahh . . .!
Were you scared?
Of course I'm scared! Mum! Dad! [laughs]. I'm thinking of my father and
    Mum and . . . ahrr.
What were you most scared of?
Capsize – I put my . . . I think I have my life jacket on all the time. You don't
    know when the ship goes.

(Interview with Seafarer H)

It seems as though sound and jiggling or moving sensations after a storm are
recalled earlier than visual recollections. These memories last long after a storm
has passed. For Ingold (2011, 133), maritime 'landscape' and weather blend into
each other. In what he describes as 'haptic engagement', bodies, materials and
land are sewn into each other 'along the pathways of sensory involvement'
(Ingold, 2011, 133). As explained earlier, a ship and those sailing on it will
become enmeshed in this maritime 'landscape' and woven into the patterns that
weather and seascape play with each other. Such patterns are also described by
Bissell as combining patterns of 'affective and embodied experience of mobility'
(2010, 485). If seafarers might have been exposed to only mildly irritating
vibrations during times of relatively calm weather and stability, once the weather
develops into a storm, all senses and materials – sea, ship and seafarer – are in
turmoil. Seafarer P, tears welling up in his eyes, described his feelings when he
faced a storm, and realised that the ship's engine was not working properly:

For example, when we [recently] were facing a storm . . . if you experience
    that [he laughs] maybe you are crying . . . Because you don't know what will
    happen to you.
You felt life threatened and that the storm could kill you?
Yeah! All of us! Because we were already standing by for an engine not
    working, because it [went] too slowly prrrrhhh! [makes a noise].
Oh, the engine didn't work?
Yeah! Too much damage in the ship. Containers get [moves his hands].
Containers got cut.
Yeah. Cut. And then the cargo containers get [wiggles his hands].
All mixed up.
Yeah, yeah. All control systems – most are like hit by a big truck. – Yeah,
    yeah. You cannot cope. You cannot.
Hmm – and the kitchen (galley)?
I don't know in the kitchen [laughter]. In my cabin it's like crrrrrrt! [shows].
Your cabin! Everything is like . . .
Yeah! [waves his arms].
The furniture.
Yeah.

Really? The furniture is pudumpudumpudum ... all through the cabin?

Yeah! Yeah! I cannot sleep because it's like pawoooo! [moves his arms wide] and you can feel the ship like crrrrrr!

Oh my goodness! Paahdoom! How long did this take, did this last?

One day. One day. But it started during the night.

Did you feel seasick?

I've already overcome seasickness.

Even under these conditions?

Yeah. I just only tashoooww, crrreewwwee [shows how he was dealing with the mess] ... Yeah. All of my things, papers, everywhere. Something you think ... something get wet and [rubs his hands].

Oh my god.

Yeah! [laughs] That's the worst thing for a seaman ... Yeah. A situation like this you have to accept.

(Interview with Seafarer P)

During a massive storm, as described above, containership technology and materiality is challenged. Vannini and Taggart (2014) explain how materials transform as a result of moving and intersecting with one another, and compare the transformations of water with those of life. For seafarers who face the extremes of a typhoon, even a giant containership with its apparently indestructible technology will momentarily appear fragile, and there is a fear that one 'cannot cope'. The lines within the meshwork of the seafaring experience muddle into a knot in which physical survival becomes more important than the economy. Escaping disaster is an immediate experience that temporarily involves corporeal and sensual surrender and focused activity. Putting the materiality of the ship back together in the aftermath is, just then, secondary. Seafarer P explains that after the storm took hold, it was not possible to continue as usual:

We cannot go inside [the port], because no pilot will go on the ship because the sea is still very rough. We go turning around the area, waiting for the pilot.

That must have caused a long delay.

Yeah, yeah. That caused ship delay. Too much. Three days maybe?

After this experience did you have time to stay still for a while to do repairs?

No, we sailed already after [this port]. Only the cargo's been removed and some cargo could not be removed because the container is almost ... it's kaput. [They] had to leave it, it's too much.

Just imagine somebody's furniture was in this container and got completely crashed.

Yeah!

(Interview with Seafarer P)

The shipping industry is only efficient as long as ships keep moving. Delays caused by disasters have to be dealt with as quickly as possible. Repairs will be made on the go. And emotional repairs will also have to be made while moving on. The events described above show how the oceans are intertwined with the everyday life of people (Anderson and Peters, 2014, 3), through the meshwork of ship/life/sea. A delay or destruction of cargo because of unexpected, violent events at sea will have consequences for the assemblage of shipping lines, cargo management and ultimately consumers. Nodes need to be adjusted and reconnected. To keep ship and cargo on schedule, one, two or more ports might have to be left out this time around the loop. Customers at this end of the line will have to wait.

Keeping a ship moving after a storm also has consequences for the everyday life of seafarers as they will have to keep dealing with the destruction of their cabins, galley and the ship's technology. The ship is in need of being 're-balanced' or settling into a harmonious state. Extra repair and maintenance may be needed on the ship's engine and other equipment. Possibly extra shifts will be required and will reduce shore leave. This loop will become narrower. Monotonous day- and night-time activities might be interrupted for a while, and then intensify again. Daily life will be accompanied again by the vibrations of the ship, with spontaneous sudden swells. The journey continues; another knot in the meshwork will be left behind.

Conclusion

Being ensounded and to bejiggled by the sea and the ship's materialities is not the usual focus for a paper on seafarers, the invisible heroes of the economy. Without the shipping industry there would be a very different everyday life for most people living in capitalist systems on shore (Urry, 2014). There is a tendency in economic structures to overlook the sensualities and corporealities involved with producing these systems. There is a tendency to overlook the very realities of seafarers for 'living with the seas'. Reflecting on some of the thoughts and experiences shared at the beginning of this paper, I consider that spaces are always under construction and, drawing on Ingold, that movement and knowledge are intertwined and constituted along lines and paths:

> Life is lived … along paths, not just in places, and paths are lines of a sort. It is along paths that people grow into a knowledge of the world around them, and describe this world in the stories they tell.
>
> (Ingold, 2007, 2)

Enmeshed in these lines and paths are the histories, materialities, languages and sensualities of those that embark on them. In this chapter, by way of considering seafarers as sensuous as well as economic beings, I have attempted to contrast Ingold's concept of unruly lines (leaning on Deleuze and Guattari's rhizomes), as can be found in *wayfinding*, with those vectoral lines of *transport*, to outline what it is to be alive with the sea, not in a recreational capacity but in a working capacity. Any incident experienced during work and life on ships, whether it is the fairly smooth monotony of taking up daily tasks, or whether it is dealing with sudden swells, is accompanied with a need for homeostasis. Such balancing or rebalancing requires not only good steering, keeping the engines going, and keeping the ship on an even keel. It includes the need for finding ways of dealing with restlessness and fatigue, with the underlying possibility of disaster, and with the power hierarchies present on ship.

References

Alderton, T., Bloor, M., and Kahvesi, E. (2004). *The Global Seafarer. Living and Working Conditions in a Globalized Industry.* Geneva: International Labour Office.

Anderson, B. (2009). Affective atmospheres. *Emotion, Space and Society*, 2(2), 77–81.

Anderson, J., and Peters, K. (2014). Introduction: 'A perfect and absolute bank': Human geographies of the ocean. In J. Anderson and K. Peters (Eds.), Waterworlds: Human geographies of the ocean. Ashgate, Farnham, pp. 3–22.

Anim-Addo, A., Hasty, W., and Peters, K. (2015). The mobilities of ships and shipped mobilities. *Mobilities*, 9(3), 337–349.

Ashmore, P. (2013). Slowing down mobilities: Passengering on an inter-war ocean liner. *Mobilities*, 8(4), 595–611.

Attlee, J. (2011). *Nocturne: A Journey in Search of Moonlight.* London: Hamish Hamilton.

Birtchnell, T., and Urry, J. (2015). The mobilities and post-mobilities of cargo. *Consumption Markets and Culture*, 18(1), 25–38.

Bissell, D. (2009). Travelling vulnerabilities: Mobile timespace of quiescence. *Cultural Geographies*, 16(4), 427–445.

Bissell, D. (2010). Vibrating materialities; mobility–body–technology relations. *Area*, 42(4), 479–486.

Borovnik, M. (2005). Seafarers' 'maritime culture' and the 'I-Kiribati way of life': The formation of flexible identities. *Singapore Journal of Tropical Geography*, 26(2), 132–150.

Borovnik, M. (2011a). The mobilities, immobilities and moorings of work-life on cargo ships. *Sites*, 9(1), 59–78.

Borovnik, M. (2011b). Occupational health and safety of merchant seafarers from Kiribati and Tuvalu. *Asia Pacific Viewpoint*, 52(3), 333–346.

Borovnik, M. (2017). Night-time navigating: Moving a containership through darkness. *Transfers*, 7(3), 38–55.

Brown, M. (2015). Seascapes. In M. Brown and B. Humberstone (Eds.), *Seascapes: Shaped by the Sea*. Farnham: Ashgate, pp. 13–26.

Cook, M., and Edensor, T. (2017). Cycling through dark spaces: Apprehending landscape otherwise. *Mobilities*, 12(1), 1–19.

Deleuze, G., and Guattari, F. (1987). *A Thousand Plateaus. Capitalism and Schizophrenia.* Minneapolis: The University of Minnesota Press.

Edensor, T. (2013). Reconnecting with darkness: Gloomy landscapes, lightless places. *Social and Cultural Geography*, 14(4), 446–465.

Edensor, T. (2015). Introduction to geographies of darkness. *Cultural Geographies*, 22(4), 559–565.

Fajardo, K. B. (2011). *Filipino Crosscurrents: Oceanographies of Seafaring, Masculinities, and Globalization.* Minneapolis: The University of Minnesota Press.

Foucault, M. (1994). Different spaces. In J. D. Faubion (Ed.), *Michel Foucault Aesthetics. Essential Works of Foucault 1954–1984 (volume 2)*. London: Penguin, pp. 175–185.

George, R. (2013). *Ninety Percent of Everything: Inside Shipping, the Invisible Industry that puts Clothes on Your Back, Gas in your Car, and Food on your Plate*. New York: Holt.

Hunt, R. (2018). On sawing a loaf: Living simply and skilfully in hut and bothy. *Cultural Geographies*, 25(1), 71–89.

Hutchins, E. (1996). *Cognition in the Wild*. Cambridge, MA: MIT Press.

Ingold, T. (2000). *The Perception on the Environment. Essays in Livelihood, Dwelling and Skill*. London: Routledge.

Ingold, T. (2004). Culture on the ground: The world perceived through the feet. *Journal of Material Culture*, 9(3), 315–340.

Ingold, T. (2007). *Lines. A Brief History*. Abingdon: Routledge.

Ingold, T. (2011). *Being Alive. Essays on Movement, Knowledge and Description*. London: Routledge.

Ingold, T. (2015). *The Life of Lines*. Abingdon: Routledge.

Langewiesche, W. (2005). *The Outlaw Sea. A World of Freedom, Chaos, and Crime*. New York: North Point Press.

Martin, C. (2011). Fog-bound: Aerial space and the elemental entanglements of body-*with*-the world. *Environment and Planning D: Society and Space*, 29, 454–468.

Martin, C. (2013). Shipping container mobilities, seamless compatibility, and the global surface of logistical integration. *Environment and Planning A*, 45, 1021–1036.

Mellbye, A., and Carter, T. (2017). Seafarers' depression and suicide. *International Maritime Health*, 68(2), 108–114.

Parker, M. (2013). Containerisation: Moving things and boxing ideas. *Mobilities*, 8(3), 368–387.

Peters, K. (2015). Drifting towards mobilities at sea. *Transactions of the Institute of British Geographers*, 40, 262–272.

Pink, S. (2015). *Doing Sensory Enthnography*. Los Angeles: Sage.

Rigg, J. (2007). *An Everyday Geography of the Global South*. London: Routledge.

Sampson, H. (2003). Transnational drifters or hyperspace dwellers: An explorative of the lives of Filipino seafarers aboard and ashore. *Ethnic and Racial Studies*, 26(2), 253–277.

Sampson, H. (2013). *International Seafarers and Transnationalism in the Twenty-First Century*. Manchester: Manchester University Press.

Squire, R. (2015). Immobilising and containing: Entrapment in the container economy. In T. Birtchnell, S. Savitzky and J. Urry (Eds.), *Cargomobilities: Moving Materials in a Global Age*. New York: Routledge, pp. 106–124.

Stanley, J. (2008). Co-venturing consumers "travel back": Ships' stewardesses and their female passengers, 1919–55. *Mobilities*, 3(3), 437–454.

Steinberg, P. E. (2001). *The Social Construction of the Ocean*. Cambridge: Cambridge University Press.

Steinberg, P., and Peters, K. (2015). Wet ontologies, fluid spaces: Giving depth to volume through oceanic thinking. *Environment and Planning D: Society and Space*, 33, 247–264.

Thomas, M. (2003). *Lost at Sea and Lost at Home: The Predicament of Seafaring Families*. Cardiff: Seafarers International Research Centre.

Urry, J. (2014). *Offshoring*. Cambridge: Polity Press.

Vannini, P., and Taggart, J. (2014). The day we drove on the ocean (and lived to tell the tale about it): of deltas, ice roads, waterscapes and other meshworks. In J. Anderson and K. Peters (Eds.), *Water Worlds: Human Geographies of the Ocean*. Farnham: Ashgate, pp. 89–102.

Whatmore, S. (2002). *Hybrid Geographies. Natures, Cultures, Spaces*. London: Sage.

10 Surfers and leisure

'Freedom' to surf? Contested spaces on the coast

Easkey Britton, Rebecca Olive and Belinda Wheaton

I often see, hear and read people connecting surfing with freedom:

> *Surfing makes me feel free.*
> *Surfing can be described by one word: freedom.*
> *Surfing is freedom.*

And I understand where they're coming from. Surfing, going surfing, wave riding is a wide space of choice and movement and rhythms and experience and connection and physicality unconstrained by the explicit organisation and rules that tend to define playing sports. Surfing takes you places, asks you questions, opens up possibilities and shows you things you didn't expect. Surfing connects you back to yourself through oceanic ideas, challenges and experiences, and describing all of this as freedom certainly makes poetic sense (Olive, 2009, n.p.; italics original).

While surfers express feelings of freedom, often they appear to think less about their ability to access and experience this freedom: their capacity to be at the beach and in the water and to experience the obsessive and all-consuming relationship that can organise or even dictate their time, relationships, lifestyles and sense of being in the world. The relationships and lifestyles typically experienced by surfers have been famously captured in global surf company Billabong's slogan: 'Only a surfer knows the feeling'. Yet, as feminists have long argued, for many individuals such leisure freedoms are relative and contingent, underpinned by inequalities and injustices (Green et al., 1989). Leisure lifestyles are not freely chosen; our opportunities and access to leisure lifestyles like surfing continue to be underpinned by a range of structural, ideological and cultural constraints (Mansfield et al., 2017). Increasing scholarship about surfing has exposed the complex ways various aspects of identity, including sex/gender, constrain our freedom (Evers, 2004, 2009; Olive, 2016; Olive, McCuaig and Phillips, 2015), sexuality (Roy, 2013; Roy and Caudwell, 2014), race and ethnicity (McGloin, 2005, 2006, 2009; Walker, 2008, 2011; Wheaton, 2013) and 'local' relationships to place (Daskalos, 2007; Evers, 2009; Olive, 2015a; Schiebel, 1995). This work has highlighted that access to, and experiences of, surfing and the beach are shaped and constrained by a range of social, cultural, historical and geographical factors. Thus, here we approach freedom as a socially constructed principle that is contextually dependent, and differently experienced.

Rooted in this body of literature, which interrogates the cultural politics of surfing spaces, in this chapter we critically consider what it means to have the freedom to surf. To do so we draw on three distinct case studies that each highlight very different experiences of surfing: women's recreational surfing in Australia; the emergent surf culture of Iran; and the historic struggle of African-American surfers in California. Our case studies rely on a mix of ethnographic fieldwork, interviews and, underscoring this, auto-ethnographic feminist reflexive practice. In each case, we were talking and surfing with people who are often marginalised, as surfers and within local and global society more broadly. As white, educated, middle-class, able-bodied female researchers and activists, we occupied positions of relative power in the spaces we worked in, even as we were, or could be, marginalised in those spaces ourselves. Such reflexivity and critical reflection have been developed and explored in feminist literature for several decades. Feminist researchers have contributed concepts of 'strong reflexivity' (Harding, 1996), 'positionality' (Rose, 1997), 'situated knowledges' (Haraway, 1991) and 'thinking the social through the self' (Probyn, 1993, 3), as part of a feminist critique of mainstream scientific methods that privilege masculine notions of objectivity, scientific detachment and value neutrality. Today, reflexivity is commonly used to refer to a method that all qualitative researchers 'can and should use to legitimize, validate, and question research practices and representations' (Pillow, 2003, 175).

This chapter is not an exploration of the challenges of positionality and anxieties we faced in our research, but we do wish to acknowledge that we cannot know the effects of the positions we occupy/occupied in relation to our participants and collaborators. What we can be clearer about are the positions we occupied in relation to our analysis and to each other. As white, middle-class, educated, able-bodied women from Anglo-Celtic backgrounds (Éire/Ireland, Australia and England respectively), our analysis emerges from numerous subjective connections. Similarly, our theoretical and activist commitment to a feminist approach to our work (see, for example, Ahmed, 2017; Hooks, 1981; Mansfield et al., 2017) creates further links and connections that are clear in this chapter. Much more difficult is accounting for the intersectional nature of our own subjectivities in relation to those of our participants and the cultures and spaces we were researching, a task we have not attempted here, but which we have variously engaged with elsewhere (see, for example, Olive and Thorpe, 2011; Wheaton, 2002).

In what follows, we individually examine experiences of freedom whilst surfing as shaped by multiple aspects of subjectivities, some of which were easy for us to access and make sense of, while others continue to elude us. In each case outlined, we occupied different and shifting positions as an insider or outsider, in terms of identity, culture, geography and community. For example, working in Iran, Easkey had distance in relation to the socio-cultural, geographical and colonial aspects of the lives of the women she was working alongside.

However, her identity as a woman who surfs opened up possibilities for the connection to occur. For Rebecca, her identity as a 'local' member of the surfing community where she was doing fieldwork impacted who she could access and posed challenges for critically interrogating her community and home. Yet this same intimacy allowed for richness and trust in her interviews and fieldwork. For Belinda, assumptions about her (white) ethnicity and nationality (Britishness) impacted the willingness of surfers of colour in the USA to speak with her, but often in unexpected ways (see Wheaton, 2013). Following Donna Haraway (1991) and Sandra Harding (1991), Gillian Rose (1997) reminds us that 'all knowledge is marked by its origins', and thus is never universal. Instead there is great value in 'knowledges that are limited, specific and partial' (Rose, 1997, 307), which highlights the diversity of lived knowledges and experiences.

Each of the case studies presented in this chapter makes limited, specific and partial claims, and so, in turn, says something of relevance specifically to the people who participated. However, placed alongside each other as they are, the experiences can also tell us something collectively. Speaking across subjective, spatial, cultural and temporal experiences and understandings, the diverse contexts help us to tease out some shared and unique ways surfers experienced freedom. Drawing on these comparisons, we suggest that while dominant surfing culture defines the beach as a free and fun-orientated space and constructs it as a place of transcendence and freedom, we also need to recognise the beach as an important site where bodies and subjectivities are created and are policed (see Evers, 2008; Khamis, 2010; Burdsey, 2013). This is fundamental because too often the struggle for identity, belonging and acceptance of those who are not the somatic white male norm still remains hidden and obscured in most dominant surfing spaces. As Lewis has argued, surfing's utopianism, particularly as imagined in surf media, and reproduced through the surf industry, ultimately ignores and denies these 'hegemonies and injustices that constrain the shoreline' (Lewis, 2003, 70). Understanding the complexities of accessing and experiencing the beach, ocean and surf is key to dismantling hegemonies, addressing injustices, and diversifying how the beach is used and by whom.

Rebecca Olive, 'women's recreational surfing: negotiating cultures of localism'

Australia's beaches are key to mainstream Australian national identity (Booth, 2001a). They mark our continental edges and acted as the first point of contact for colonial encounter. Smashed ships, lost lives, held hands, arrivals, last looks, political protest, sex and pleasure, violence and death, a place of harvest and nutrition, a place to watch the sun rise and set – beaches have long shaped everyday life for people across the continent now known as Australia. Surfing waves has long been a part of beach life: indigenous Australians have played in the waves for millennia, and since the early 1900s stand-up surfing has become a key part of Australia's beach cultures (Osmond, 2011). Yet not surprisingly, not all surfers are considered equal, so which surfers hold the power when it comes to accessing beaches, surf breaks and lineups? This question has been central to my work, focusing on the challenges facing women who surf face in contemporary surfing culture, linking this to localism and how it shapes who has rights to access beach spaces (Olive, 2015b, 2016; Olive, McCuaig and Phillips, 2015).

The intimate connections surfers develop to surf breaks are often framed in terms of 'being local'. People who claim local status claim to know 'their' place best, 'from an inside that is spoken for, and for which the locals have the right to speak and act. The right ... comes through a long association and knowledge of a place, its names and its history' (Evers, 2009, 127). This can include historical, geographical and cultural knowledges such how a wave breaks, where the rocks are, how the tides affect a break, memories of epic swells and of surfers past (Evers, 2009). Being local to a surfbreak can cut across ability and affords the surfer priority over waves (see also Daskalos, 2007; Preston-Whyte, 2002; Schiebel, 1995). Nick Carroll, a renowned Australian surfer and surf journalist writes that many surfers 'feel a sense of ownership' over surf breaks 'that makes land-based property rights seem feeble in comparison' (Carroll, 2000, 60). In Australia, this way of thinking disrespects the spatial relationships held by Aboriginal Australians (Evers, 2008; Garbutt, 2011), while at the same time using settler histories to position more recent immigrant Australians as outsiders. While this tension is not my focus here, it is impossible to ignore how surf localism continues to reflect and reproduce colonial legacies and practices of white settlers, and how this links to contemporary surfing contestations over surf-spaces (see also the third section of this chapter).

Coupled with the often dominating place that surfing takes in a surfer's identity, the intense sense of locally based 'rights' and 'ownership' over a surfbreak can create feelings that 'Newcomers ... threaten to trouble the places locals have quietened and made their own' (Garbutt, 2011, 127). Building on the work of Margaret Henderson (2001, 329), who described surfing as a 'fantasised last frontier for anxious men and youths', I have found that women are still considered newcomers to surfbreaks and surfing culture. Localism is therefore a key part of Australian surfing culture, with deep-rooted relationships to place for some, while excluding the possibility of the same relationships for others (Evers, 2008). Surfing localism is cultural, developed through male-dominated traditions, media and hierarchies, but this is shifting and the contributions of women – and other newcomers – to surfing culture are being recognised as continuing to change the culture of surfing itself.

Women, surf media and voices of authority

Historically, men have controlled representations of surfing and surfers in Australia, and in this way have controlled the culture and practice of surfing itself. Although women have long been participants, recreational surfing culture and its media has systematically marginalised women. Women surfers have had less media and industry representation and win less prize money, while being hyper-sexualised in the media, with their participation minimised in the water (Henderson, 1999, Henderson, 2001; lisahunter, 2016; Olive, 2014; Phillips and Osmond, 2015).

For example, of the 38 inductees into the Australian Surfing Hall of Fame only 6 are women and not all women's world title holders are inducted. Surfer Layne Beachley was only inducted in 2009, two years after winning her seventh world title, with Stephanie Gilmore inducted in 2013; the year she won her fifth world title. The book *The Ninth Wave*, published in 2011 by respected magazine *Australian Surfing World*, presents '100 amazing surfing images', coupled with the story of each image. In this book, not a single woman was included – as a surfer, a storyteller or a photographer (Olive, 2011). Yet out in the surf, in a growing number of coastal places, women are a commonplace and growing presence. Women who surf are active in representing and supporting themselves and other women on social media, and their participation is challenging how we think about who is local and dominant in the surf. More recently these shifts have been reflected in some surf media. In 2016, Australian surf magazine *White Horses* published an all-women issue, and *Surfing World* hosted surfer Lauren Hill and her partner Dave Rastovich as co-editors. An increasing number of boutique and niche magazines, films and surfwear brands are being launched by women, and the images of surf photographers, such as Kristen Scholtz, Clare Pluekhahn, Fran Miller and Ming Nomchong, are regularly featured across surfing's mainstream and social media. However, like surfbreaks, the cultures of surf media operate differently depending on the context of the publication, and while these developments are exciting, they remain notable. For example, male photographers continue to dominate, and all mainstream magazines are still edited by men.

The continuing dominance of men as the established 'locals' of surfing, can be understood by what Nikolas Rose (1996) calls a 'Voice of Authority', which is a vital component in the ways we constitute ourselves as surfers and locals. This voice is ever-present in local graffiti, and non-local city-based surfers are well acquainted with the effects of such authority (Olive, 2015a). Voices of authority help us see how and why a sense of freedom is available, or not available, for various people in certain surf spaces. If we can only become a voice of authority through traditions established by white, high-performance, short-boarding men, how is it possible for newcomers – especially non-male, non-white, inexperienced surfers – to feel a sense of belonging in the surf?

How this is evident in everyday, recreational surfing was clear in research I did in Byron Bay, a small town which is a popular surfing and tourist destination in northern New South Wales on Australia's east coast. I interviewed a number of women who live and surf in Byron Bay, between 2009–2011, to discover how they understand and experience male-dominated cultures of surfing (Olive, 2013). The ways the women I spoke with felt excluded in the water may be a result of explicit violence, or more subtle cultural assumptions which are difficult to locate. Despite the violence of some men, such as in Sophie's story below, every woman I spoke with in my research felt that she was supported by the men in her surfing community. However, they also recognised that even with their skill and the support afforded them, women are not able to take a 'natural' place in the lineup or to get respect, in the way that men are.

SOPHIE: I've been in the water when women will never get back in the water again [...] Because they've had a really bad experience with, a man who's been quite violent [...] and it's enough for them to go, 'You know what? I don't want to do this anymore. I'm too scared' [...] I remember one girl [...] who was just starting to learn to surf [...] She wasn't in his way, but he just went ballistic [...] 'What are you doin' in the water [...] you shouldn't be surfing, and go and lie on the beach with all the other girls' [...] And he's violent. I've seen him go off at another girl [...] 'What are you doin'? You should have an apron on. You should be at home, cooking!'

(Interview, 2009)

GEORGIE: ... just coming down to that it's not just your surfing. Like, that a girl could surf better than them, but she won't take that natural place in the lineup out there [...] I s'pose there's guys that would say that that happens to them out there as well. I'm sure it does [...] I just think it would probably happen to a girl more.

(Interview, 2009)

Since they could often be excluded from key take-off spots in the water, the women I spoke with adopted an ethical approach by surfing down the line, away from practices of hassling, dropping and snaking that they didn't want to be part of. They developed excellent knowledge about how waves form and break at their regular breaks, and were able to get lots and lots of waves. For many of these women, this was an ethical practice that allowed them to avoid emulating the behaviour of the men who excluded them.

SOPHIE: it still happens to me frequently, in a crowded surf blokes have said to me, 'You've had five waves'. And I think well, yeah right [...] I've sat and I've got crummy bloody little waves and yeah! They've evolved to something else, but I've worked bloody hard to get them [...] But they've held me into account for the waves that I've had [...] I will find those small little windows and I'll be happy with that. I'll get a hundred waves, don't you worry. I will find them and I will get them.

(Interview, 2009)

Nonetheless, their position down the line could lead to moments of patronising behaviour, where men offer unsolicited advice, or, as has happened to me on more than one occasion, giving an un-needed 'little push' into a wave (Olive, McCuaig and Phillips, 2015). The advice – and the little push – might be offered with good intentions, but the effect diminishes women's capabilities. Some women subsequently reproduce the competitive practices of male surfers, but most women, despite feeling patronised, manage to tolerate such behaviour with compassion and patience. By adopting ethical approaches to the line-up, women are actively contributing to cultural change in the surf. The wave knowledge they develop makes them more visible as skilled surfers, and their less aggressive approaches become new voices of authority amongst surfers – women and men – who respect such ways of being in the sea.

ABI: Okay, that, this is my pet peeve. If you're one of like, two women out, and the other girl drops in on you? I, I really, I really, that's a pet peeve for me. And, it's like sometimes I think they do, the other girl will drop in on you just because they want to show, like, 'I'm a better surfer here' or 'this is my break' or whatever! But [. . .] you need to be like, 'Yes! Come out here whenever you can!' [. . .] Like there'll definitely sometimes be camaraderie because, you're both girls. You don't even have to know each other, you're like 'Yeah! Go!' and you'll yell and cheer for each other or you'll give each other a smile.

(Interview, 2009)

SOPHIE: Well, you know, you just gotta be determined. You know, just try and stay really calm and be really rational about it and, try not to, you know, cross over to the dark side really. I mean, once, once you, walk those, those steps, once you start to be like them then, you're like them then aren't you.

(Interview, 2009)

Despite what the dominant men might believe, not everyone gives them authority because of their surfing ability. For me, and for the women I interviewed, some of the most admirable surfers are those who are generous in spirit, who watch out for others and let waves pass them by rather than taking them all just because they can. My own surfing ethics have developed in response to the voices of authority I admired as I learned to surf, which were primarily the voices of women, using the men as a reference point of what to avoid. And while I definitely don't always succeed in following that, I always turn back to them as my point of reference. Similar things are happening on Instagram, where women are subverting the historical bikini-clad, beach-bound images of women by posting images that show them in the water and promote their skills as surfers; that emphasise their knowledge of coastal places and cultures; and, most wonderfully, that celebrate the role other women who surf play in their lives (Olive, 2015b).

In Byron Bay, as in many other places, women who surf are a growing visible presence and their participation is changing surfing culture. In part, they're doing this simply by continuing to paddle out every day, as well as more active efforts. Yet although the activism by women should continue to drive this kind of change and openness, a number of surf researchers, including Doug Booth (2001b), Clifton Evers (2004) and Ford and Brown (2006), emphasise the role of men in the success of cultural changes. Until men who surf respect that there can be approaches to surfing that are defined by other ways of knowing, women, and many others, will continue to struggle to be accepted and valued within surfing culture, and to make effective contributions to surfing culture. Freedom in the surf is not contingent on men, but it does remain relative to them.

In contrast to Australia, Easkey describes the different freedoms and constraints that have been negotiated by some women in Iran.

Easkey Britton, 'the emergent surf culture of Iran: negotiating social hierarchies'

I made several visits to Iran between 2013 and 2016. My role was to observe and support the collaborative potential of informal, community-based activities and to document the potential for new ways of learning and doing surfing (Britton, 2018). I did so by sharing and constructing stories grounded in lived experiences through a mix of active participation and participant observation (Malone, 1999). Despite an established lifestyle sports culture in Iran, surfing is a recent phenomenon. This section highlights the factors influencing the emergence of this surf-space, including the impact of shifting power dynamics as a lifestyle sport emerges and is formalised (Wheaton, 2013). These factors also influence uneven impacts or benefits that are experienced by various social groups of 'women who surf', who are too often represented as a homogenous whole in mainstream surf media (Olive, 2016).

As I have written in greater detail elsewhere, my first journey to Iran was in 2010 with French film-maker Marion Poizeau (Britton, 2018), when we documented our experience of surfing in Chabahar in the Sistan-Baluchistan province of south-eastern Iran, the only region in the country exposed to open-ocean swells and surf.[1] It is home to the Baloch, a traditionally semi-nomadic ethnic minority group with a long history of political unrest, persecution and marginalisation that goes largely unreported (McLoughlin, 2015). It is highly conservative and has some of the highest levels of social deprivation in the country. Considered an area of low-intensity conflict and insurgency since 2004, Mogelsen in the *New York Times* (2012) called it the 'scariest little corner on earth'. The province is largely ignored by Iran's elite classes, especially from the capital Tehran, and until recently a policy of underdevelopment persisted. This region was not a foreign tourist destination, nor was it a place commonly visited by Iranian tourists either. But we did find waves to surf. The reasons why surfing had not previously taken hold had a lot to do with the aforementioned geo-politics as well as geography. The 'surf zone' has a very small window of 80km of exposed coastline; it's difficult and costly to get to (a two and half hour flight from the capital Tehran); and a lack of resources and the trade embargo makes it challenging for even affluent Iranians to access surf equipment.

In 2012, a short film of scenes where I am covered in a hijab (required under Iranian law for all women in public space) surfing in an unusual desert landscape in Iran was released on YouTube. It was quickly picked up and shared among

Tehran's urban youth and several young female snowboarders contacted me and Marion directly. With insider knowledge and official permission, we decided to return to Chabahar in 2013 to document the experience of Iran's first recorded female surfers, two young women in particular, Mona from Tehran, a snowboarder, and Shahla from Zahedan, a swimmer who had learned how to dive in Chabahar. Considering the policy of male–female segregation in sport since the 1979 Islamic revolution, we didn't know how surfing in public spaces might be received. However, on the day we stood on the edge of the sea on Ramin beach, 6km outside the town of Chabahar, watching Mona and Shahla catch some of their first waves; applause erupted from spectators on the shore. This was the first time that everyone there had witnessed surfing. And the first surfers they saw surfing were women. 'Is surfing something boys can do too?' a local boy asked, watching these women in the water.

Since then, surfing has begun to develop rapidly. This was enabled initially through a collaborative process of co-organising surfing workshops with Mona and Shahla and their network of snowboard and diving friends, as well as with members of the local community at what has now become the primary 'surfing beach' at Ramin. Those who participate in surfing represent a diverse racial/ ethnic and gendered mix. As well as the training of female surf instructors, local surfers have trained others to teach surfing, self-organising surf sessions and establishing the Ramin Surf Club in 2016. As with lifestyle sports in other countries (Thorpe and Ahmad, 2015; McKay, 2016; Olive, 2015b), in Iran social media is an important tool for how young Iranians access and share information. However, in Iran, where much media access is blocked and regulated, Instagram (the only international social media platform that is not officially blocked), is significant in allowing young people independence in promoting these activities and representing their own experiences. Informal grassroots groups on Instagram such as *We Surf in Iran* and *Boarders of Hope* were created by some of the first female surfers in Iran, maintaining a strong visual representation of surfing as something women do. Considering the history of women in surfing (Thompson, 2015; Walker, 2011; first section of the chapter), and the issues so many women face in the Middle East, the emergence of surfing in Iran through women is, importantly, a key factor in the acceptance of surfing as 'something boys can do too'. A high proportion of surfers at surf workshops are female (Britton, 2018). These women represented a diverse mix, although the majority were urban middle-to-upper class Iranian; in their twenties, and with university-level education and/or a professional occupation. In an NBC news report a young Iranian female surfer described her experience and how surfing changed her perspective of the sea-space:

> I was terrified of the sea and hated this area but when I saw the clip of these girls surfing I became very interested and joined their workshop [. . .] I feel like I am flying – out there on the water you don't think about any of your problems.
>
> (NBC, 2016)

Despite the uptake of surfing by women, these activities remain largely inaccessible for those from lower income backgrounds and rural areas, as well as for minority groups, highlighting that for Baloch women there exist multiple inequalities of class, ethnicity, geography and gender. Although perhaps amplified in this context, the issue of access for women from minority groups prevails across cultures more generally for women in sport (Koshoedo et al., 2015). Until 2015, female participation from the local Baloch community in surfing was almost non-existent, and the few who participated were pre-adolescence.

Although some women and girls do swim, they do so in full dress, which, once wet, is incredibly heavy and cumbersome to move in. While it is possible to discuss the issue as a 'dualistic and over-simplistic notion of de-veiling-as-liberation' (Britton, 2018), the award-winning documentary of Iran's first female surfers, *Into the Sea* (directed by Marion Poizeau), highlighted the 'surf hijab' (full covering of the female body, including the head) as a key factor for the acceptance of women in surfing and their ability to actively participate. In the West, the notion of clothing as 'freedom' persists, with strong associations of veiling with oppression in Western media feeding a discourse of an Orientalist and hegemonic vision of Muslims, and Muslim women in particular (Bullock, 2002; 123; Jafri, 1998; Said, 1989). Yet Iranian scholar Hoodfar (1992, 15) argues that the veil has multiple meanings across cultures, time and space. For example, there are contrasting narratives of, for example, the success of female Muslim athletes in the 2016 Olympic Games, including Iran's first female Olympic medallist Kimiah Alizedah, and the so-called 'Burkini ban' on French beaches, forcing women to remove their hijabs (BBC, 2016; *The Guardian*, 2016). As Mona Seraji argues, clothing is not freedom, '[f]reedom is not about what you wear, it's about choosing what to do with your life' (Huck, 2015). The need for specialised sports clothing has attracted the support and interest of independent, female-owned companies such as Dutch company Capsters. They design sports hijabs for women, and women's surf-wear label Salt Gypsy's surf leggings, designed initially for sun protection and modesty, are being used in the predominantly Islamic Maldives. However, a lack of suitable sportswear for women in watersports remains a challenge. Shahla Yasini, who has continued to surf every season since 2013, adds that it prevents women freely moving and feeling comfortable in the water: '[It can] stop women not used to sport from taking their chance to get close to water' (Huck, 2018).

Another factor in the issue of access for women to surfing in Iran has been its informal beginnings in the country. One such example, which has since become known as the Be Like Water (BLW) programme, was co-created in collaboration with Iranian women and piloted in 2015 and 2016. The programme co-evolved as a way to address perceived barriers for women of mixed social class and ethnic background who wanted to participate in surfing, especially Baloch women (Britton, 2018). It focused on creating a safe and trusting space for female participants to develop a positive and personal relationship with their environment. As one participant shared: 'Surfing can showcase the beauty in our differences by allowing us to be truly who we are when we surrender to the playfulness of waves and wave-riding' (Interview, Shirin, 2016).

Surfing's play-like character seems to make it particularly open to new modes of 'doing' and 'being', in turn embracing a diverse mix of people. This realisation is not limited to surfing and within the wider sport-for-development discourse there is a shift happening with a greater 'de-sportisation' of physical activity being called for (Sterchele, 2015). Sterchele (2015), for example, highlights the importance of the non-competitive qualities of doing sport and the 'play-for-development' trend, especially when social cohesion is desired.

Some of the more intangible outcomes and impacts of the recent introduction of surfing has been a social mixing, challenging more dominant social hierarchies and a history and culture of oppression, blurring experiences of gender, class, ethnicity and geography (Roy, 2013). Surfing has opened up spaces that didn't exist before (Roy, 2013), and has changed the meaning attached to the beach-space at Ramin, shifting identities and relationships. For example, visitors are now coming to learn surfing at the local surf club. Collaboration between women from different cultures are documenting the rise of surfing in a region previously considered too dangerous to visit (Huck, 2018). Young women are discovering a new relationship with the ocean and how it shapes their sense of self. As one woman recently wrote to me: 'I finally understood the most important skill to surf, connection with mother ocean. Believe it or not [it] changed my whole being' (Shahla, August 2017).

These qualities of surfing allow for a non-prescriptive approach and learning to surf has brought other skills, such as greater ocean awareness. For example, I was sent images of a self-organised beach clean-up at Ramin, along with this comment from one of Iran's new surfers: 'Water is a multi-dimensional inter-active mirror reflecting and affecting not only our physical but also spiritual attributes' (Masood, Chabahar, 2015).

Although surfing may appear to level the playing field, connecting a diversity of people through a shared experience, it doesn't exist in a vacuum and overlays deeper power dynamics in the most underprivileged part of the country. Accessibility and entry into surfing is different for males and females, with greater limitations for female participants in terms of clothing, creating a 'safe space', and social acceptance especially for women and girls from more conservative backgrounds.

Wheaton (2013) argues that surfing is open to different interpretations, and initial self-organising mechanisms in Iran highlight a tension between informal and formal approaches. It is too early to say if the informal qualities of surfing as a lifestyle sport will continue to support this inclusivity, especially in light of recent moves to formalise the sport with the establishment of a national governing body and Iran's recognition as the 100th member of the International Surfing Association (ISA, 2016). Furthermore, with sanctions lifting, Western, commercial lifestyle sport brands are keen to market to new niche audiences.

The development of surfing in Iran needs to consider not only the gender and class differences but the interplay and potential trade-offs between the economic development of surfing as a leisure and tourism activity in a socially deprived area and the preservation of culture. As one local surfer said, 'We want to surf but we also want to maintain our culture [...] the people with money have the power' (Conversation, May 2016). Although surfing has swiftly taken hold, 'it will take time, decades perhaps, before Baluchistan gets accustomed to new trends and ideas' (Abed, local surf club manager, in Huck, 2018, n.p.). Ethnicity and social class play a huge part in accessibility to surfing and influences the power dynamics with diverging interests and motivations for how surfing could be used as a tool to foster inclusion and greater freedom.

Belinda Wheaton, 'the California beach, race and space: negotiating relative freedoms'

The intersectionality of gender, race, ethnicity, economic privilege, history and culture also underpins this final case study, which draws on research exploring the experiences of African-American surfers in California (following Wheaton, 2013). It vividly illustrates that the image of the beach as a benign, free and fun-orientated space reflects a white Western privileged position and construction. The research illustrates that the California beach was, and remains, an important site where racial formations are created and mobilities policed (see Evers, 2008; Khamis, 2010; Burdsey, 2013). While the 'black' surfers I interviewed had managed to negotiate a path to find their own place and space both at the California beach and within a surfing culture, these were often relative and at times contested spaces and freedoms. For many other African Americans in California the beach was perceived or experienced as a place where they were outsiders, even 'space invaders' (Puwar, 2004). For many, this was not a space where they felt a sense of belonging or ownership. Surfers I interviewed recognised that they were in the minority and that many black families tended not to go to the beach:

> I wouldn't call it a white thing, but we don't do it so regularly. Black families that are, quite 'water enlightened' do. But they're still a minority. You know, the majority of people that live in the inner city don't use the facilities of the beach. They don't feel for whatever reason either comfortable because they're brought into it, or because there's not enough black people down there to allow them to feel comfortable
>
> (Interview, 2011)

While a range of important socio-cultural, economic, ideological and logistical factors have contributed to this pattern, 'the historical and cultural patterns of oppression', which have subsequently become systematically embedded in 'society's norms and daily practices' (Erickson et al., 2009, 531), are most central.

From the turn of the twentieth century through to the 1920s, the southern California beach was an important leisure space for the African-American community as well as for other minority groups in California. As Jefferson's research outlines, the beach was a place for 'relaxation, recreation and vacation' (Jefferson, 2009, 156). However, when segregation was legally imposed in the 1920s, the city beaches around Los Angeles, as elsewhere, were typically limited to whites only. For example, in Santa Monica, which had a diverse ethic community, just one small area (at the foot of Pico Boulevard) remained accessible for, and popular with, the African-American community. It was marked 'Negroes only' and became known as the *Inkwell* (Jefferson, 2009).

While the racial restrictions in public spaces in California was legally invalidated in 1927, 'de facto segregation' continued through to the 1960s (Jefferson, 2009, 155). Additionally, practices of discrimination and harassment were adopted to 'discourage' African Americans from visiting, using facilities or settling in particular beach locales (Jefferson, 2009, 180). For example, African Americans were barred from the private beach clubs emerging on the Santa Monica coastal strip with fenced-in beach to keep the 'undesirables' out, and white landowners added real estate covenants with 'Caucasian restrictions' on their properties, which had a significant – and still enduring – impact on home ownership (Jefferson, 2009). A philosophy was emerging across southern California beach towns that beaches should be 'reserved for the white public', typified by the 'Save the beaches for the public' campaigns emerging during the 1920s. These campaigns were created initially to stop African Americans creating their own beach clubs (Jefferson, 2009, 180). As Jefferson outlines, 'refusal to allow African Americans access to various places of leisure constituted an informal policy' along many stretches of the California coastline, which was vigorously 'enforced by many white citizens and policy makers' (Jefferson, 2009, 156).

The impact of this historically driven spatialisation of race continues into the present time (Cresswell, 1996). Factors including economic segregation via gentrification and white postcolonial anxiety have fuelled what Stenger describes as the 'flight to and fortification' of white suburban spaces by the white middle classes (Stenger, 2008, 44). Minority ethnic groups tend to congregate in city centres and the California beach suburbs, like many other coastal areas in the United States and internationally (Burdsey, 2011), tend to have overwhelmingly majority white populations. Statistics confirm that people living along the Los Angeles coastline are 'disproportionately non-Hispanic, white and wealthy, compared to the state and county: sixty-eight percent are non-Hispanic white, and less than five percent are black. In all coastal communities, the black population was too small to be significant' (USC Coastal democratic study, cited in Garcia and Baltodano, 2005, 195). In the most affluent beachfront enclaves like Malibu, private ownership of rows of beachfront property has made access to the beach almost impossible other than for the homeowners. Some have sought to completely cut off non-residents' beach access, for example posting illegal 'private property' signs, and employing private security guards (Project, 2010). These beachside spaces continue to be constructed through 'imaginary notions of whiteness' (Nayak, 2010, 2375), operating as places of 'white retreat and safety' (Neal, 2002, 446) that still includes practices of white territoriality.

A second related factor contributing to this racialisation of leisure space is the long-standing historically rooted perceptions of outdoor nature-based spaces as 'white' (Erickson et al., 2009) held by both African Americans and white communities. As research on BME (black and minority ethnic) usage of various outdoor leisure spaces in the USA confirms, nature-based (but often free) recreational resources, from beaches to mountains, remain the 'purview of the White middle-class visitor' (Wolch and Zhang, 2004, 416). Visiting nature was not seen as part of black culture (Wheaton, 2013). As Erickson et al.'s (2009, 538) research outlines, for African Americans in the US specifically, fears exist about travelling outside of their (usually urban) 'comfort zone', based on both perceptions about being in certain spaces and historically rooted anxieties about personal safety. They discuss how the woods in particular continue to have a 'strong negative connotation' because of their association with poverty and lynching (Erickson et al., 2009, 539). Over time, therefore, nature and the outdoors in general become viewed as places to be avoided, which has been passed down through family and community, impacting the leisure patterns of many African Americans. As Erickson et al. outline, and my research with the African-American surfers also confirms, 'because of how individuals have been taught to think about natural areas, invisible lines of segregation continue to exist' (Erickson et al., 2009, 543). As one of the surfers I interviewed explained:

> So it's kind of been like a, my personal mission to let folks know, to let black people know, so that we can open the door to new opportunities and new thinking to think differently about ourselves so that we can take advantage of all the things that might be available to us.
>
> (Interview, 2009)

Furthermore, because 'recreating in natural areas' is not considered a 'black thing' to do, those who choose to take part risk 'being associated with 'white culture', which can be perceived by other African Americans as rejecting 'black culture' (Erickson et al., 2009, 540). Similarly, African-American surfers spoke of the difficulties they faced because surfing was seen as 'acting white' (Wheaton, 2013.

This case study reminds us that people from different racial and ethnic groups construct 'different meanings for natural spaces', including beaches, based on their own values, cultures, histories and traditions (Garcia and Baltodano, 2005, 197). Even amongst this small community of black surfers, freedom to surf was contingent and variable, based not just on race but the intersection with gender, class and geography. Nonetheless, for many, experiencing freedom took on a different set of meanings to those perpetuated in the popular discourses of surfing. Just to claim space and visibility at the beach, whether to picnic, play or surf, involves transcending long-standing historically rooted perceptions of space and culture. As one surfer put it: 'I believe that we are engaged in a civil rights struggle at the beach. That is much different than what white surfers are doing' (Interview, 2009). At times this was a battle – as one interviewee put it, 'we surfed to experience the freedom and in a group we were sometimes willing to fight for that freedom' (Interview, 2009). However, as activist groups, like Los Angeles-based The City Project (Project, 2010, 145), whose mission is to achieve equal justice, democracy and liveability for all of the community, recognise, public access to the beach is integral to equality, with 'people of colour' and those on low incomes 'disproportionately denied the benefits of coastal access'. The (California) beach therefore represents an essential 'front in the struggle for equal justice'.

Conclusions

This chapter has engaged with individual notions and experiences of 'freedom' in relation to how we 'live with the seas' as surfers. Our differing research about surfing and the beach has, together, explored how freedom is contingent and relative to intersections across race, sex/gender, sexuality, ethnicity, history, culture and geography. In all of this, bodies are key as the site for understanding our lived experiences of freedom, and for experiencing the operation of power in any context: space, place, culture or time.

Surfing has retained an 'assumed innocence both as an imagined space and as it physically manifests itself as a place removed from the everyday concerns of power, inequality, struggle and ideology' (Carrington, 2010, 4).[2] It has therefore been seen to offer a 'space for transcendence' and 'utopian dreaming', a liminal space that is distinct from seemingly more significant areas of 'civic life' (Carrington, 2010, 4). Yet, as our case studies have illustrated, beaches are contested sites where power is exercised. As Garcia and Baltodano argue,

> Beaches are not a luxury. Beaches are a public space that provide a different set of rhythms to renew public life. Beaches are a democratic commons that bring people together as equals ... public access to the beach is integral to democracy and equality.
>
> (Garcia and Baltodano, 2005, 145)

In this chapter, we have illustrated how a sense of freedom at the beach and in the surf is contingent upon a complex and locally specific set of intersectional factors. However, how these factors impact on individuals is also dependent on the constitution of their own subjectivity, and how they are able to negotiate these diverse constraints. Recognising these power relations that shape the discursive construction of the beach as a 'utopia' and the sea as 'free' is an essential step in understanding how people can, and do, experience coastal leisure generally, and surfing specifically.

Our chapter illustrates that when we think through some of the dominant discourses about the seascape – such as freedom, and what this means – it is always contextual, and negotiated. Assumptions about and feelings of 'freedom' have been revealed as historically key to discourses in surfing culture, yet are much more complex and contested for users who do not, will not, or cannot fit dominant surfing norms. However, for those who have experienced exclusion, this sense of freedom can be profound in that it is fought for and thus not taken for granted. As several African-American surfers articulated, freedom at the beach was central to wider claims for recognition, for equality and for identity.' 'I don't see the vision of black people surfing as separate from black peoples' liberation globally' (Black Surfing Association member, cited in *A Soul Surfer's Quest*).[3]

Notes

1 Surfing has been recorded in the Caspian Sea as an occasional occurrence for some time, possibly introduced when the US military were stationed there.
2 Carrington writes here not about surfing but sport, but his comments are apposite in this context too.
3 From, http://www.youtube.com/watch?v=CZH-Up1Kk-g&feature=related (Accessed 21 February 2017)

References

Ahmed, S. (2017). *Living a Feminist Life*. Durham: Duke University Press.
BBC. (2016). Liberté, Egalité, Burkini? [Online] http://www.bbc.com/news/blogs-trending-37176299 (Accessed 24 May 2018).
Booth, D. (2001a). *Australian Beach Cultures: The History of Sun, Sand and Surf*. London: Frank Cass.
Booth, D. (2001b). From bikinis to boardshorts: *Wahines* and the paradoxes of surfing culture. *Journal of Sporting History*, 28(1), 3–22.
Britton, E. (2018). 'Be like water': Reflections on strategies developing cross-cultural programs for women, surfing and social good. In M. Mansfield, J. Caudwell, R. Watson and B. Wheaton (Eds.), *The Palgrave Handbook of Feminism and Sport, Leisure and Physical Education*. London: Palgrave Macmillan, pp. 793–807.
Bullock, K. (2002). *Rethinking Muslim Women and the Veil: Challenging Historical and Modern Stereotypes* IIIT.
Burdsey, D. (2011). Strangers on the shore? Racialized representation, identity and in/visibilities of whiteness at the English seaside. *Cultural Sociology*, 11(5), 537–552.
Burdsey, D. (2013). 'The foreignness is still quite visible in this town': Multiculture, marginality and prejudice at the English seaside. *Patterns of Prejudice*, 7(2), 95–116.
Carrington, B. (2010). *Race, Sport and Politics: The Sporting Black Diaspora*. London: Sage.
Carroll, N. (2000). Defending the faith. In N. Young (Ed.), *Surf Rage: A Surfer's Guide to Turning Negatives into Positives*. Angourie, NSW: Nymboida Press, pp. 54–73.
Cresswell, T. (1996). *In Place/Out of Place: Geography, Ideology, and Transgression*. Minneapolis: University of Minnesota Press.
Daskalos, C. (2007). Locals only! The impact of modernity on a local surfing context. *Sociological Perspectives*, 50(1), 155–173.
Erickson, B., Johnson, C. and Kivel, B. (2009). Rocky Mountain National Park: History and culture as factors in African-American park visitation. *Journal of Leisure Research*, 41(4), 529–545.
Evers, C. (2004). Men who surf. *Cultural Studies Review*, 10(1), 27–41.

Evers, C. (2008). The Cronulla riots: Safety maps on an Australian beach. *South Atlantic Quarterly*, 107(2), 411–429.

Evers, C. (2009). 'The Point': Surfing, geography and a sensual life of men and masculinity on the Gold Coast, Australia. *Social and Cultural Geography*, 10(8), 893–908.

Ford, N. and Brown, D. (2006). *Surfing and Social Theory: Experience, Embodiment, and Narrative of the Dream Glide*. London: Routledge.

Garbutt, R. (2011). *The Locals*. Bern: Peter Lang.

Garcia, R. and Baltodano, E. F. (2005). Free the beach! Public access, equal justice, and the California coast. *Stanford Journal of Civil Rights and Civil Liberties*, 143. [Online]. https://www.cityprojectca.org/ourwork/documents/StanfordFreetheBeach.pdf (Accessed 12 September 2011).

Green, E., Hebron, S. and Woodward, D. (1989). *Women's Leisure, What Leisure?* Basingstoke: Macmillan.

Haraway, D. (1991). *Simians, Cyborgs, and Women: The Reinvention of Women*. New York: Routledge.

Harding, S. (1991). *Whose Science? Whose Knowledge?: Thinking from Women's Lives*. New York: Cornell University Press.

Harding, S. (1996). Standpoint epistemology (a feminist version): How social disadvantage creates epistemic advantage. In S. P. Turner (Ed.), *Social Theory and Sociology: The Classics and Beyond*. Oxford: Blackwell, pp. 146–160.

Henderson, M. (1999). Some tales of two mags: Sports magazines as glossy reservoirs of male fantasy. *Journal of Australian studies*, 62, 64–75.

Henderson, M. (2001). A shifting line up: Men, women and *Tracks* surfing magazine. *Continuum: Journal of Media and Cultural Studies*, 15(3),319–332.

Hoodfar, H. (1992). The veil in their minds and on our heads: The persistence of colonial images of Muslim women. *Resources for Feminist Research*, 22(3/4), 5.

Hooks, B. (1981). *Ain't I a Woman: Black Women and Feminism*. Boston: South End Press.

Huck. (2015). The Iranian female snowboarder challenging perceptions of gender in the Middle East: Meet Mona Seraji. [Online] http://www.huckmagazine.com/ride/snow-2/mona-seraji/ (Accessed 24 May 2018).

Huck. (2018). Shahla Yasini is leading a surf revolution in Iran: Riding a wave. [Online] http://www.huckmagazine.com/outdoor/surf/ramin-surf-iran/ (Accessed 24 May 2018).

ISA (2016). *International Surfing Association Adds Iran As Landmark 100th Member*. News Bulletin. [Online] https://www.isasurf.org/international-surfing-association-adds-iran-as-landmark-100th-member/(Accessed 21 February 2018)

Jafri, G. J. (1998). The portrayal of Muslim women in Canadian mainstream media: A community-based analysis. *Report of the Afghan Women's Organization, Toronto*. Canada.

Jefferson, A. R. (2009). African American leisure space in Santa Monica: The beach sometimes know as the 'Inkwell', 1900–1960s. *Southern California Quarterly*. 2 August, 155–189.

Khamis, S. (2010). Braving the Burquini: Re-branding the Australian beach. *Cultural Geographies*, 17(3), 379–390.

Koshoedo, S. A., Paul-Ebhohimhen, V. A., Jepson, R. G. and Watson, M. C. (2015). Understanding the complex interplay of barriers to physical activity amongst black and minority ethnic groups in the United Kingdom: A qualitative synthesis using meta-ethnography. *BMC Public Health*, 15(1), 643.

Lewis, J. (2003). In search of the postmodern surfer: Territory, terror and masculinity. In J. Skinner, K. Gilbert and A. Edwards (Eds.), *Some Like It Hot: The Beach as a Cultural Dimension*. Oxford: Meyer and Meyer Sport, pp. 58–76.

lisahunter. (2016). Becoming visible: Visual narratives of 'Female' as a political position in surfing: The history, perpetuation, and disruption of patriocolonial pedagogies. In H. Thorpe and R. Olive (Eds.), *Women in Action Sport Cultures: Identity, Politics and Experience*. London: Palgrave Macmillan, pp. 319–348.

Malone, K. (1999). Environmental education researchers as environmental activists. *Environmental Education Research*, 5(2), 163–177.

Mansfield, M., Caudwell, J., Watson, R. and Wheaton, B. (2017). Introduction: Feminist thinking, politics and practice in sport, leisure and physical education. In M. Mansfield, J. Caudwell, R. Watson and B. Wheaton (Eds.), *The Palgrave Handbook of Feminism and Sport, Leisure and Physical Education*. London: Palgrave Macmillan, pp. 1–16.

McGloin, C. (2005). *Surfing Nation(s) – Surfing Country(s)*. Unpublished PhD, School of Sciences, Media and Communications, Faculty of Arts, University of Wollongong.

McGloin, C. (2006). Aboriginal culture: Reinstating culture and country. *International Journal of Humanities*, 4(1), 93–99.

McGloin, C. (2009). Nation, country and indigenous surfing. *Kurungabaa: A Journal of Literature, History and Ideas from the Sea*, 2(1), 87–91.

McKay, S. (2016). Carving out space in the action sports media landscape: The skirtboarders' blog as a 'Skatefeminist' project. In H. Thorpe and R. Olive (Eds.), *Women in Action Sport Cultures: Identity, Politics and Experience*. London: Palgrave Macmillan, pp. 301–318.

McLoughlin, S. (2015). *Mass Atrocities, Risk and Resilience: Rethinking Prevention*. Leiden: Brill-Niehoff.

Mogelson, L. (2012). The scariest little corner of the world. *The New York Times*. [Online] https://www.nytimes.com/2012/10/21/magazine/the-corner-where-afghanistan-iran-and-pakistan-meet.html?pagewanted=3&_r=0&seid=auto&smid=tw-nytmag (Accessed 22 May 2018).

NBC. (2016). Iranian women stoke the surfing revolution. [Online] https://www.nbcnews.com/news/world/iranian-women-stoke-surfing-revolution-n628091 (Accessed 22 May 2018).

Nayak, A. (2010). Race, affect, and emotion: young people, racism, and graffiti in the postcolonial English suburbs. *Environment and Planning A*, 42(10), 2370–2392.

Neal, S. (2002). Rural landscapes, representations and racism: examining multicultural citizenship and policy-making in the English countryside. *Ethnic and Racial Studies*, 25 (3), 442–461.

Olive, R. (2009). Freedom by any other name. Blog post on *Making Friends with the Neighbours*. 18May. [Online] http://makingfriendswiththeneighbours.blogspot.co.uk/2009/05/i-often-see-hear-and-read-people.html (Accessed 21 February 2018).

Olive, R. (2011). The Ninth Wave: No girls allowed. Blog post on *Kurungabaa: A Journal of Literature, History and Ideas from the Sea*, 21 March. [Online] http://kurungabaa.net/2011/03/21/the-ninth-wave-no-girls-allowed/(Accessed 21 February 2018).

Olive, R. (2013). *Blurred Lines: Women, Subjectivities and Surfing*. Unpublished PhD, School of Human Movement Studies, The University of Queensland.

Olive, R. (2014). Imagining surfer girls: The production of Australian surfing histories, *Girl Museum*. [Online] http://www.girlmuseum.org/wp-content/uploads/2014/12/History-Rebecca-Olive-Imaginging-Surfer-Girls.pdf (Accessed 21 February 2018).

Olive, R. (2015a). Surfing, localism, place-based pedagogies, and ecological sensibilities in Australia. In B. Humberstone, H. Prince and K. Henderson (Eds.), *International Handbook of Outdoor Studies*. London: Routledge, pp. 501–510.

Olive, R. (2015b). Reframing surfing: Physical culture in online spaces. *Media International Australia, Incorporating Culture and Policy*, 155, 99–107.

Olive, R. (2016). Going surfing/doing research: Learning how to negotiate cultural politics from women who surf. *Continuum: Journal of Media & Cultural Studies*, 30 (2), 171–182.

Olive, R., McCuaig, L. and Phillips, M. G. (2015). Women's recreational surfing: A patronising experience. *Sport, Education and Society*, 20(2), 258–276.

Olive, R. and Thorpe, H. (2011). Negotiating the F-word in the field: Doing feminist ethnography in action sport cultures. *Sociology of Sport Journal*, 28(4), 421–440.

Osmond, G. (2011). Myth-making in Australian sport history: Re-evaluating Duke Kahanamoku's contribution to surfing. *Australian Historical Studies*, 42(2), 260–276.

Phillips, M. G. and Osmond, G. (2015). Australia's women surfers: History, methodology and the digital humanities. *Australian Historical Studies*, 46(2), 285–303.

Pillow, W. (2003). Confession, catharsis, or cure? Rethinking the uses of reflexivity as methodological power in qualitative research. *International Journal of Qualitative Studies in Education*, 16(2), 175–196.

Preston-Whyte, R. (2002). Constructions of surfing space at Durban, South Africa. *Tourism Geographies*, 4(3), 307–328.

Probyn, E. (1993). *Sexing the Self: Gendered Positions in Cultural Studies*. London: Routledge.

Project. (2010). Bruce's Beach: The City Project celebrates Black History Month. In *The City Project Blog*. Los Angeles. [Online] https://www.cityprojectca.org/bruces-beach (Accessed 21 February 2018).

Puwar, N. (2004). *Space Invaders: Race, Gender and Bodies out of Place*. Oxford: Berg.

Rose, G. (1997). Situating knowledges: Positionality, reflexivities and other tactics. *Progress in Human Geography*, 21(3), 305–320.

Rose, N. (1996). Identity, genealogy, history. In S. Hall and P. Du Gay (Eds.), *Questions of Cultural Identity*. London: Sage, pp. 128–150.

Roy, G. (2013). Women in wetsuits: Revolting bodies in lesbian surf culture. *Journal of Lesbian Studies*, 17(3–4), 329–343.

Roy, G. and Caudwell, J. (2014). Women and surfing spaces in Newquay, UK. In J. Hargreaves and E. Anderson (Eds.), *Routledge Handbook of Sport, Gender and Sexuality*. London: Routledge, pp. 235–244.

Said, E. W. (1989). Representing the colonized: Anthropology's interlocutors. *Critical Inquiry*, 15(2), 205–225.

Schiebel, D. (1995). Making waves with Burke: Surf Nazi culture and the rhetoric of localism. *Western Journal of Communication*, 59(4), 253–257.

Stenger, J. (2008). Mapping the beach: Beach movies, exploitation film and geographies of whiteness. In D. Burnardi (Ed.), *The Persistence of Whiteness: Race and Contemporary Hollywood Cinema*. London: Routledge, pp. 28–50.

Sterchele, D. (2015). De-sportizing physical activity: From sport-for-development to play-for-development. *European Journal for Sport and Society*, 12(1), 97–120.

Thompson, G. (2015). *Surfing, Gender and Politics: Identity and Society in the History of South African Surfing Culture in the Twentieth-Century*. Unpublished PhD, Stellenbosch: Stellenbosch University.

Thorpe, H. and Ahmad, N. (2015). Youth, action sports and political agency in the Middle East: Lessons from a grassroots parkour group in Gaza. *International Review for the Sociology of Sport*, 50(6), 678–704.

Walker, I. H. (2008). Hui nalu, beachboys, and the surfing boarder-lands of Hawai'i. *The Contemporary Pacific*, 20(1), 89–113.

Walker, I. H. (2011). *Waves of Resistance: Surfing and History in Twentieth-Century Hawaii*. Hawaii: University of Hawaii Press.

Wheaton, B. (2002). Babes on the beach, women in the surf: Researching gender, power and difference in the windsurfing culture. In J. Sugden and A. Tomlinson (Eds.), *Power Games: A Critical Sociology of Sport*. London: Routledge, pp. 240–266.

Wheaton, B. (2013). *The Cultural Politics of Lifestyle Sport*. London: Routledge.

Williams, M. (2007). *Legendary Surfers: Nick Gabaldon (1927–1951)*. [Online] http://www.legendarysurfers.com/2005/02/black-surfer-nick-gabaldon.html (Accessed 21 February 2018).

Wolch, J. and Zhang, J. (2004). Beach recreation, cultural diversity and attitudes to nature. *Journal of Leisure Research*, 36(3), 414–443.

11 Students and teachers

Te hone moana/the ocean swell: learning to live with the sea

David Irwin

The influential Millennium Ecosystem Assessment (MEA) (UNESCO, 2004) found that 60 per cent of the planet's ecosystems were degraded, and that no place on the planet was free from human impact. Significantly, the MEA also found that species were becoming extinct approximately 1,000 times faster than fossil records indicated had occurred in the past, with much extinction the result of lost habitat or over-exploitation. Environmental scientists have used the term Anthropocene (the human epoch) to mark this geological period of mass extinction that rivals the great extinctions in the past (for example, see Whitehead, 2014). Nothing like this has happened in over 65 million years and yet it continues to unfold beyond the notice of most people (Hamilton, 2010).

Our planet is a watery one. The oceans cover approximately 70 per cent of the earth's surface, and because of the vastness of these watery spaces, they have seemed infinite and unalterable by human effort. However, on a global scale, the oceans are under extreme stress. For example, decades of over-fishing have seriously threatened the biological diversity of the oceans to the point where many fisheries have already collapsed (UNESCO, 2004). How this has occurred is no secret, for, as Elliot explains,

> every hour of every day, humans are removing nearly 10,000 tonnes of fish from the ocean – these numbers are almost unfathomable, but when they are extrapolated into an annual catch the number seems even more staggering: ninety million tonnes of fish and other marine life are harvested from the world's oceans every year.
>
> (Elliot, 2014, 249)

The extensive research into the state of global fish stocks conducted by the MEA found that 90 per cent of all large predatory fish were gone (UNESCO, 2004). This loss of oceanic life has significantly reduced the biological diversity of the oceans to the point where many marine species and ecosystems are threatened. However, biological diversity is just one of the planetary systems under threat from human activity.

Reflecting on the sobering reality described by the MEA, I remember when I was first introduced to the concept of planetary system boundaries. Rockström et al. (2009) used this term to describe the parameters within which the predictable functions of key planetary systems occur. These boundaries are based upon the stable planetary conditions of the Holocene period that comprises the last 12,000 years. It is these stable conditions that have allowed humans to settle, agriculture to flourish, and other species to adapt to niches relevant to particular habitats. Nine planetary system boundaries have been identified: boundaries for climate change, stratospheric ozone depletion, atmospheric aerosol loading, ocean acidification, biochemical flows of nitrogen and phosphorous, fresh water use, land systems change, biosphere integrity and chemical pollution. Rockström et al. (2009) describe how the safe boundaries allowing life as we know it to flourish have been breached across numerous systems, first into a zone of increasing risk, where the authors argue that caution should be employed, and then into a dangerous zone of possible irreversible unpredictability and potential collapse.

This analysis is both disturbing and challenging. As an educator, prior to learning about planetary boundaries my key message to students had been about understanding the circumstances described by the MEA, and encouraging behavioural change to help *prevent* catastrophic systems collapses. But the argument presented by Rockström et al. (2009) emphasises that collapse is already taking place across biological diversity systems (where species are rapidly becoming extinct at an accelerated rate both on land and in the oceans); land use systems (where widespread alteration of landscapes for purposes such as intensive agriculture has significantly changed or destroyed large-scale biological systems including important marine ecosystems such as mangrove and estuarine landscapes); ocean acidification (where the changing acidity of the oceans as a result of radically increased atmospheric carbon dioxide is impacting on ocean ecosystem integrity); and the application of chemical fertilisers releasing nitrogen and phosphorous that have corrupted natural chemical system balances both on land and in the ocean. According to Rockström et al. (2009), these systems are either about to or already have entered an unpredictable phase of variability that lies outside of the mostly stable conditions of the last 12,000 years. Climate change, regarded as a system also in crisis, is likely to increase the unpredictability of the other systems, especially biological diversity systems. This is because climate change disrupts the integrity of habitats and exacerbates extinction rates for species that are no longer adapted to the place where they have lived.

What I found most challenging was how to process the information presented by Rockström et al. (2009). What did this information mean for my teaching? Encouraging students to take action to save the oceans and the planet offered hope to students. But what does an educator do with the very real possibility that no matter what actions are taken, the planet may well have already passed into unprecedented cycles of realignment driven by human activity, a process that would see (from a human perspective) the irreversible collapse of the life-supporting systems we and much other life on the planet depend on. How do educators engage students with these crises in a meaningful way, and what should they try to achieve through that engagement?

This chapter is a personal reflection on teaching and learning about these issues. It is about engaging with the coast as a local place and exploring one's relationship with it. In Aotearoa New Zealand (Aotearoa NZ) this process aligns with the Māori tradition of turangawaewae (expressed as a deep sense of belonging), which acknowledges the need for a place to stand coupled with an obligation to care for that place. It includes learning about other life forms that reside in the coastal waters and the interconnections that exist; and, most importantly, beginning to unravel the complexity of the challenges that confront current generations of humans, challenges that no other generation has faced before. The chapter reflects on four key aspects of teaching and learning that I have come to understand as critical to effective pedagogy in the twenty-first century that may apply not only in Aotearoa NZ, but to any educator engaging with students in respect of how they might live with the sea.

First, students need to understand the issues at stake on both global and local scales. Second, they need to develop their action competence (Jensen and Schnack, 1997) so that they can explore behaviour change on their own terms. Third, students must understand the need to adapt to a changing world; and finally they need to learn to value living in the moment. These dimensions of teaching and learning help to develop concepts of identity (how the students see themselves) and community (understanding their social context) which are central to changing human attitudes and behaviours. The following discussion traces this evolving journey of realisation. Through my reflections, I recognise that identity and community are central to changing human attitudes and behaviours in relation to the ocean and to the future health of the planet.

Turangawaewae: finding out about the place where I stand

In Aotearoa NZ, no part of our small country is far from the ocean. We are an island people and many of us have a deep sense of belonging to the coastal margins of the land and to the sea, although the same may be said of many other communities around the world (for example, see D'Arcy, 2006; and also Chapters 4 and 5 in this volume). I grew up by the ocean in what was then a small coastal village and I have fond memories of a childhood spent in, on and beside the ocean. I remember as a child of about 10 or 12 going down to the beach to gather cockles for lunch. As I made my way off the beach I encountered some local council workers who were eager to share my kai moana (sea food) and they suggested I gather some for them as well, which I did. As I gave them the cockles, they remarked that I was now an honorary Māori (people indigenous to this country). I took this to be a compliment, for in the Māori world view, sharing of kai moana was a token of friendship, and gifting took place over very long distances to maintain relationships.

I also took the remark as a compliment because the ocean was (and remains for many Māori) a sacred place, as it is for most Polynesian peoples. I had learned the histories of discovery in school and also of the stories that informed those early peoples of who they were. Central to the Maori world view is the notion of turangawaewae, or a place to stand. Turangawaewae was established through family connections to place, acknowledging that the people and land/seascapes are inextricably entwined. Turangawaewae also included aspects of obligation to the place where one stood, a respect for and reciprocity with the land and associated coastal/marine areas that are key elements of many indigenous beliefs but often missing in modern world views.

Over the years as a tertiary[1] environmental and outdoor educator in Christchurch, a city perched on the edge of the Pacific Ocean, I have worked with many students of differing backgrounds. This work has always involved helping students to explore their own sense of turangawaewae, and it is always a shock to encounter students who have never or seldom seen the ocean despite living only a short distance from it. However, as Christchurch has some of the lowest socio-economic suburbs in the country, it should not have come as a surprise that for some families even a bus fare to the beach can be a burden. For the children in these families, any engagement with the ocean can be incredibly satisfying in terms of finding out about their community and their place within it.

Within the formal schooling system, getting students out of classrooms to explore their local environment seems to be increasingly problematic for a number of reasons. For example, the cost of taking students out of school is increasingly becoming an issue in Aotearoa NZ because of inadequate government funding to cover out-of-school activities (Irwin, 2015), although it should be noted that such constraints are present in many other countries as well. However, of critical importance is the trend towards a narrowing of curriculum that has been associated with ever more pervasive neoliberal agendas. For example, research into curriculum delivery in New Zealand schools undertaken by Fisher and Ussher (2014) found that schools are increasingly focused on National Standards and there is limited coverage of specific learning areas other than literacy and numeracy across school targets, professional development, budget or school priorities, suggesting a narrowing of the curriculum. The New Zealand Council for Education Research (Bolstad, Joyce and Hipkins, 2015) drew similar conclusions regarding environmental education. At the very moment in history that humans need to understand their connection to place, the educative practices that allow development of a sense of place are diminishing.[2] This is unfortunate, for, as Elliot (2014, 259) maintains, 'the next few years will be critical in making positive change for the future of our marine resources. The ocean is the land's life blood and much of it is dying'.

Against this backdrop lie stories of inspiration. A young teacher I work with is an extremely passionate educator, environmentalist and surfer who is bucking this trend. He has had the opportunity to develop year-long, one-day-a-week programmes for year 12 and 13 high school students.[3] The programme is contextualised mostly in marine space and uses achievement standards (a form of national assessment) from a variety of curriculum areas including education for sustainability; social studies and geography, and earth and space sciences while also engaging students in practical activities such as snorkelling and surfing. His reflection on the unique curriculum that he had designed included the following observation:

The issues that pertain to the coastal and marine environment, locally and globally, are often political; infused with contrasting information and vague future directions … I wanted the students to emerge with an understanding of these issues, an ability to discern political bias amongst scientific data, and the confidence to communicate these issues.

(Irwin and Brasell, 2014, 25)

This teacher observed that the course had a positive impact on the students. They became much more aware of the world around them and of the problems and issues that confront us collectively. But also, over time many of them gained more confidence being in the water; they became more proficient swimmers and as a result were more comfortable to explore their watery world. However, most importantly he observed that students were expressing a connection to the coastal and marine environments that they had learned about and engaged with through their experience with the programme. This connection was expressed in a number of ways, including wanting to care for the coastal places they visited through removing plastic and other debris from the beach.

Engaging with the ocean in different ways can have significant impacts on how students see themselves in relation to the world around them. For example, one student studying with us observed in her learning journal[4] after her first snorkeling experience that

[l]earning to snorkel was a real eye opener and it is fair to say it changed my life. I did not realise there was a whole other world under the water that is absolutely teaming with life; I had become used to only seeing the surface of the water and had never really understood what lay beneath until I took the time to look there.

Another student reflected in their journal that 'surfing taught me a love and respect for the ocean, and has helped grow my thirst for knowledge of the world around me'. Such observations can lead to quite remarkable changes in the way people think and act.

We have a course within our sustainability and outdoor education degree called Beach and Surf Education. The aim of the course is to use activities including snorkelling and surfing as hooks to engage students experientially with the increasingly significant degradation of the oceans. Within the course, students are required to think of creative ways to teach difficult and often emotionally challenging concepts to secondary school students in a meaningful way that they can relate to. In the summer following BP's Deepwater Horizon oil spill into the Gulf of Mexico in 2010, one of the students wanted to explore the notion of suffering experienced by marine mammals in the area as a result of the spill. The student took his classmates down to the water line and had each of them drizzle black acrylic paint on each other, roll in the sand and splash water on themselves. He then took photos of the students (see Figure 11.1) to be used in a presentation that followed in the nearby surf club. He alternated the images of the students smeared with paint with images of wildlife struggling with the sticky tarlike oil from the devastating spill in the Gulf of Mexico, and in so doing created a shocking reality check. The futility of trying to scrape oil from fur and feathers was brought home for the class. The utter helplessness of the animals was emphasised. In sum, the ecological devastation of the spill was established in an affective and engaging way. It is something the participants will be unlikely to forget.

Quoting Ian Rankin, the student concluded his presentation with the observation: 'You wouldn't think you could kill an ocean would you? But we'll do it one day. That's how negligent we are' (Rankin, 2008, n.p). Grappling with this sentiment, students would later discuss the disconnection of most people in the developed world with the ocean. This is a disconnection that allows us to overlook the role we all play in creating a market for fossil fuels and at the same time distances us from events such as BP's Deepwater Horizon oil spill.

An important part of the Beach and Surf Education course is to explore this disconnection in an effort to move beyond feelings of being overwhelmed by the vastness of the problems associated with the ocean. In the simplest of terms, students ask the questions 'how could people let this happen?' and 'what can we do to make things better?' Through a focus on young ocean activists such as Dave Rastovich,[5] students begin to perceive the ocean as a contested space, a place where the values of neoliberalism that allow for events such as BP's

Figure 11.1 Imagining being covered in oil. Photo by student, used with permission.

Deepwater Horizon oil spill to take place are being challenged. By learning of the actions of people such as Dave Rastovich, students come to understand the concepts that are at the heart of turangawaewae, and the ways in which obligation to place might be demonstrated. Other courses in their degree require students to become actively involved in issues with a view to finding solutions to problems and promoting change.[6] In the discussion that follows, I reflect on my own learning and teaching with students about living with the sea.

Our legacy of plastic: learning to take action

Every year I take students to a remote beach situated in a small bay on Banks Peninsula called Ikoraki. In pre-colonial times, this beach was home to a few Māori families, but the only sign remaining of their occupation are some shallow depressions in the earth where their whare (homes) were located over 170 years ago. Later in the late 1830s and early 1840s, this beach was the site of one of the most profitable shore whaling stations in Canterbury. The beach is still littered with whale bones, with more revealed with each passing year as the sea eats away at the places where they were buried. Bits of broken china and the ceramic stems of tobacco pipes can also be found, and one can still distinguish the charred soil and brickwork defining the location of the tri pots where the whale blubber was boiled down and the earth caught fire when the pots overflowed. Through this era, pregnant female southern right whales journeyed north from Antarctica and frequented these small bays to give birth and to nurse their young before moving into deeper waters where the calves matured. The whalers would row their boats out to the whales and harpoon the young calves, knowing the mothers would linger with their ailing calves, allowing an easy shot at the much larger and potentially more dangerous mother. As a result of the technique of killing two generations at once, whales are not a common sight at the bay even after all this time. Students are quick to make links between these unsustainable practices of the past with regard to hunting whales and current unsustainable practices of industrial fishing that have seen the collapse of many fish species.

Besides the evidence of past whaling activity, the beach is now also covered in large and small pieces of plastic and polystyrene detritus. Our conversations inevitably trace the impacts of humans on this beach over time and we wonder how people in the future will view the remnants of our modern developed societies and the waste we have abandoned in the oceans. I vividly remember the first time students and I came across a dead pied shag at Ikoraki, its partially decomposed body exposing a crop containing an assorted array of plastic fragments. There have been many more carcasses of birds encountered since then, and I find myself repeatedly wondering how our modern existence has come to the point that the things we take for granted are so toxic to nature. Meduna (2015) describes how plastic is now almost unavoidable for seabirds in this country as they forage for food offshore. She notes that while plastic items were found in the stomachs of less than 5 percent of seabirds in 1960, by 2010 approximately 80 percent had swallowed anything from plastic bags and bottle tops to bits of synthetic clothing that had washed out into the ocean from urban rivers, sewers and waste deposits.

(Meduna, 2015, 61)

Indeed, by 2050 there will be more plastic in the ocean than fish by weight (Wearden, 2016). I struggle with these revelations and, like many others I know, have developed the habit of retrieving plastic lying in the gutters as I walk along our local streets, ensuring one less piece of the deadly material will be swept into the ocean through the stormwater systems.

As an educator, the inundation of our oceans with plastic is a complex but very necessary issue to explore, for plastic forms the foundation of our techno-logical landscape and there is little in our lives that does not include plastic. A group of students found a novel way to demonstrate this notion when they were preparing for a class presentation they were about to give. They decided to move everything out of their flat that contained plastic or nylon onto the front lawn, and they were astounded to find that there was so little left inside the flat. They took photos of everything on the lawn and used these to provide a visual representation of how pervasive plastic is in our lives.

The same student who had explored the BP Deepwater Horizon oil spill described earlier was also very disturbed by the amount of plastic he encountered suspended in the water whilst diving at a local beach. This personal, embodied encounter heightened his awareness of the issue as well as his motivation (and obligation) to do something about it. In his words, 'this experience brought the issue closer to home. I've seen the documentaries on ocean islands of rubbish but my dive made me aware that this is really an issue for us in Christchurch' (CPIT, 2013). Focusing on plastic water bottles such as Pump manufactured by Coca-Cola, the student began a movement to ban the bottle.

The Ban the Bottle movement was instigated by the student as an action project that was a requirement of a course called Implementing Sustainable Practice. Action projects allow for the development of action competence (Jensen and Schnack, 1997), explained by Jensen as 'a process in which students identify environmental issues, determine solutions, and take actions in ways that develop their competence for future action to solve or avoid environmental problems' (Jensen, 2002 as cited by Eames et al., 2006, 8[7]). The student decided he would use his skills as a photographer to raise awareness of the issue and he explored a range of ideas before eventually deciding to photograph students with black acrylic paint dribbling from their mouths to represent the amount of oil required to make and transport one plastic bottle filled with water. He developed a series of striking posters that asked 'What is the real price of bottled water?' (See Figure 11.2). The student set up a Facebook page to engage students in the project, and to collect and make available a range of resources related to the project.

The student hoped his action project would more widely expose the problems associated with consuming bottled water in the hope of making the campus water

what is the real
price of bottled
w**a**ter?

CPIT

Figure 11.2 Ban the bottle. Photo by student, used with permission.

bottle free, just one small step towards reducing the amount of plastic he had witnessed in the ocean. When asked on local student radio why he implemented the action, he maintained there was no point in purchasing bottled water when the city's water supply was amongst the purest in the world, and therefore the behaviour was wasteful, unnecessary and contributed to the marine pollution he had witnessed. The process of being interviewed reinforced the role he had assumed as an agent of change, and the persona that developed as a result of the action was extended beyond the confines of the course. For example, he would later write an article about the excessive wastefulness of bottled water for *New Zealand Surfing Magazine* (Goodman, 2013, 89). In the article he concluded that 'it is the people with the power ... of course we have the power to change, we know and understand the consequences of our actions but choose to ignore them' (Goodman, 2013, 89).

The student arranged to meet the chief executive of Christchurch Polytechnic Institute of Technology, who gave her support to the Ban the Bottle project; she also allowed the institution's logo to be used on the posters and gave permission for them to be displayed around the campus and on social media. The chief executive also issued a warning to the company contracted to operate the campus cafés that the action project may have an impact on their sales of bottled water, and advised the executive that any new building plans or refurbishments should make adequate provision for drinking fountains.

Over the next two years, a succession of students continued with the Ban the Bottle action project. Several students were granted $6,000 by the chief executive the following year to produce eight 3m × 3m canvases of the art works that featured in an open-air art gallery adjacent to the campus, which provided further endorsement and celebration of the action. Powerful people were on board and the students quickly found they had established a community network willing to assist with the project. Emboldened by their success, that year also saw the students take Ban the Bottle to other local institutions including the University of Canterbury, the Otago University Medical School and the YMCA. At each site, management welcomed the students, and through their support extended the social action community.

This action drew together people from across different organisations and spanned several years; it carried powerful messages that were difficult to ignore. The initial action led to another and another and another, with the message relating to disposable water bottles spreading across the city; those students responsible for the action gained in confidence and clearly felt that they could make a difference to the world they lived in. Experiences with such projects are critical to students' learning, for the experience of successfully leading a change process plants the notion that change is both possible and achievable, and affects how the students see themselves. However, earlier discussion about planetary boundaries leads me to the conclusion that this learning alone is no longer adequate preparation for students as they face the future. This is because some planetary systems are already in a state of unpredictability, and change is now inevitable regardless of actions that are taken. This means the notion of action competence in preparing students to 'solve or avoid environmental problems' (Jensen, 2002, as cited by Eames et al., 2006, 8) is fraught with difficulties. The reality is that the future state of the planet is very uncertain and the ability of the planet to support the needs of future generations can no longer be taken for granted (UNESCO, 2004). In addition to building action competence capabilities, educators need to help students adjust to this unfolding reality.

The rising tide: adapting to a changing world

People in Aotearoa NZ are familiar with the power of the sea as well as with the risks of flooding and erosion that are an integral part of living near the coast. However, these risks are escalating. As the sea rises, ground water will also rise, coastal flooding will become more common, and erosion will increase. These processes will have far-reaching impacts on coastal towns and cities around the world.

My family's home is located about 500 metres from the beach in a small coastal village called Sumner. I can hear the waves breaking on the shore from my house and when conditions permit, I can ride my bike to the beach with my longboard under my arm. The village is located on very low-lying land less than one metre above spring high tide in places, although this is not unique in Christchurch. According to the Parliamentary Commissioner for the Environment (PCE), at least 9,000 other homes are also located less than 50 centimetres above spring high tide levels (2015). To put this figure in perspective, the PCE explained that this was more than the number of homes that lay within the Red

Zone following the powerful 2011 earthquake that destroyed much of the city; these homes were finally demolished, the land eventually to be returned to the forested wet-land river margins characteristic of the pre-colonial past.

The Christchurch City Council has attempted to rezone much of our village along with significant areas in Christchurch as likely to be inundated as a result of sea level rise. Not surprisingly, there has been a well-orchestrated public backlash against the rezoning from people fearing that the process had been both rushed, and that their property values would plummet as a result of the move. Taken to court by a concerned residents group that formed for the purpose of contesting the council decision, the council eventually withdrew the inundation zoning amendment to the city plan, acknowledging that greater public consultation and community engagement in the process was needed. Clearly the public were not prepared for the action taken by the council, yet it is clear that the sea level is rising and our coastal communities will increasingly be at risk. Many other people live in similar coastal environments around Aotearoa NZ and some have already lost homes and farmland. The solutions are quite challenging for those property owners at risk. Education obviously has a key role to play in the way people conceptualise and engage with these increasingly pressing problems, and assisting future educators to develop practices that encourage learners to engage with them is critical.

In an activity trying to conceptualise this issue, two students had the rest of the class build elaborate cities on the beach in a site of their group's choosing. The explanation of the activity did not mention the ocean, but rather emphasised a competition to build realistic cityscapes based on sustainable principles. The student groups constructing the cities quickly gravitated to the damp sand lower down the beach where it could be more readily sculpted and shaped than the dry sand above the high tide mark. The cities grew in complexity; office blocks, apartment towers, football fields, roads and houses, public transit systems, water catchment systems and renewable energy sources were all discussed by the groups and constructed. Eventually prizes were awarded for the most elaborate designs, and we retired inside for more theoretical discussions on sustainable and unsustainable city design (a very real subject for students in Christchurch following the 2011 earthquakes). After several hours had passed, the two students requested to continue with their activity, which the rest of the class had thought was long over.

We returned to the beach to find the incoming tide beginning to encroach on the sustainable sand cities. The obvious task now was to defend our cities against the incoming tide, which was of course a fruitless exercise. Raising infrastructure was not possible and building physical barriers only temporarily halted the rising tide, although hastily constructed sea walls permitted the cities to exist below sea level for a short while. Eventually the only option was to abandon the devastation that was taking place and relocate higher up the beach. The students who had built the cities complained to the student facilitators that they should have explained what the real aims were!

We sat on the sand dunes amongst the marram grass and watched as our cities succumbed to the incessant pounding of the incoming waves. The facilitators then led a most revealing discussion about the places we choose to live, why we choose to live there, and how little foresight we humans actually have regarding the decisions we make. We also talked about how many of the coastal places we choose to live are also on the margins of places that are vital breeding and feeding sites for many marine animals. As sea levels rise, these locations will become increasingly compressed between human coastal defences protecting against inundation and the rising ocean.

Links between human behaviour and sea level rise become very real conceptually and emotionally when you watch the city you have built slowly sink beneath the rising tide. Climate change, the health of the ocean, sea level rise and human endeavour are inextricably linked, and the city of Christchurch, built on the low-lying estuarine and river margins and drained wet-lands behind the sand dunes where we sat, now seemed incredibly vulnerable. The students correctly concluded, as the PCE had poignantly observed, 'What the world, including our small country, does now will affect how fast and how high the sea rises' (2015, 7).

As highlighted earlier in the chapter, sea level rise is but one of many challenges that we must now face. Discussion about planetary system boundaries identified a state of increasing unpredictability across multiple planetary systems, including climate change, ocean acidification, loss of biological diversity, changes in land use and flows of nitrogen and phosphorous. As these planetary systems drive change towards conditions that are unprecedented in recent history, human societies will also have to adapt to new ways of living. Adaptation will undoubtedly cause significant stress, as the example of rising sea levels and threat of inundation has done for my local community of Sumner. Yet the vulnerability of the members of our small village pales to insignificance alongside the estimated 2 billion of the planet's poorest people identified by the MEA as vulnerable to sea level rise, storm surges and risk of inundation (UNESCO, 2004). Change is not an easy process to accept, especially if change is accompanied by widespread hardship and a sense of loss. Therefore, the ability to adapt to changing circumstances coupled with strategies to develop resilience to the stress of change on both personal and community levels become vital aspects of teaching and learning.

Paradigm shift: learning to live and being in the moment

Despite having thought and educated about these matters for the last two decades, I am both personally and professionally challenged by the situation we now find ourselves in. Educators and academics alike call for a paradigm shift, a move in values, attitudes and behaviours towards more sustainable ways of living. Indeed, Milbrath (1989) was able to demonstrate evidence of this shift among 5,000 research participants across multiple countries nearly 30 years ago, and more of us have undoubtedly moved somewhat further along this spectrum since then.

However, as much as we talk about doing things differently and educate about doing things differently, the degradation of the oceans and the planet as a whole continues at an increasingly alarming rate. Even the most educated people continue to live lifestyles that are destructive to the planet and all life upon it. This is because sustainable ecological footprints (the ecological spaces we occupy), are almost impossible to achieve in the developed world if we embrace the values and behaviours promoted in that space. With students it is important to draw this discussion down to a personal level, for, as Rahmstorf and Richardson explain,

> we still have a choice between a reckless experiment with an unprecedented rapid rise in CO_2 concentration and a 'slash and burn' style exploitation of marine resources on the one hand, and a sustainable future with healthy oceans on the other. The fight over this choice of direction is fought in many arenas – and it is affected by the choices that all of us make every day, when we buy a car or refrigerator, when we plan our holiday, or simply when we talk to friends about these issues
>
> (Rahmstorf and Richardson, 2009, 223)

Importantly, when we can encourage students to think of themselves as part of larger communities and movements comprised of people making similar decisions based on what is best for the planet, they are more likely to take actions such as Ban the Bottle, as discussed earlier. This is because social change occurs in communities and generates feelings of togetherness, thus avoiding the feeling that they acting alone in their efforts to change (Della Porta and Diani, 2006).

However, according to Rahmstorf and Richardson,

> we cannot achieve [a sustainable future] through a multitude of individual adjustments. We are facing the challenge of critical fundamental questioning of our lifestyle and consumption and patterns of production. We must grapple with the complexity of the entire earth system in a forward-looking and precautionary manner.
>
> (Rahmstorf and Richardson, 2009, xi)

Of course, Rahmstorf and Richardson are referring to a paradigm shift, but what is often overlooked is the uncomfortable reality that paradigms do not tend to shift incrementally, but rather in sudden collapses driven by events of a size and magnitude that are exceedingly disruptive to human populations (Capra, 1982). Wars, famines, diseases, extreme weather events, water and food shortages resulting in conflict and tidal inundation of cities have all been historical drivers precipitating paradigm shifts (Wright, 2004; Diamond, 2005). As discussed earlier, Rockström et al. (2009) have also identified ecological systems now at risk of unpredictability and irreversable collapse, adding into the mix other more permanent planetary changes driving paradigm shift.

As educators we need to remain cognisant of this reality because if we convey the belief that paradigm shift can be achieved incrementally we are almost certainly not being realistic and risk alienating a generation of learners from the reality of social change. But at the same time we must encourage a new way of thinking and behaving, creating a sense of turangiwaewae (belonging and obligation), concepts that are critical foundations for the future. But how do we keep students engaged and passionate about the world around them when there is so much bad news?

Known as the Shark Man, surfer and ecologist Riley Elliot speaks of being passionate about life, following your loves and enjoying life for what it has to give. It is through such engagement he finds the will to continue to fight for the well-being of the oceans and, in his particular case, sharks (Elliot, 2014). He suggests that people 'not dwell on the past or future, for you will negate the beauty of the now' (Elliot, 2014, 275). This is an important concept for students to grasp. We find ourselves in our present predicament because of decisions that have been made by many generations that have preceded us. Therefore, the burden of responsibility is intergenerational and must not be allowed to rest with the current generation of young people. Students must learn to embrace the moment, to find value in their lives. Xavier Rudd is a musician popular with my students and his music is highly politicised. In his song *Follow the Sun*, Rudd (2012) speaks of the weight of knowing about the problems confronting us, and about remembering your place by the ocean. As a surfer, Rudd finds solace and a sense of turangawaewae in the waves, and I encourage students to do the same. We need to reinforce the connection to the place where we stand, and through that strengthened connection to place and community, find resilience to face the changes we are confronted with.

Parting thoughts: te hone moana/the ocean swell

This chapter has traversed what I have reflected upon as being the key elements of teaching and learning to engage students in the issues of our time. Students need to: understand the issues at stake; develop their ability to take action; develop and adapt to a changing world; and learn to value living in the moment. In this way, concepts of identity (how the students see themselves) and community (understanding their social context) are central to changing human attitudes to the ocean and to the future health of the planet.

In Aotearoa NZ, the influence of the ocean has significantly shaped who we are and how we live, and will continue to do so into the future, for we are a maritime people living by, on and with the sea. However, this is not unique to Aotearoa NZ. On a local and global scale people are shaping the oceans; radically altering delicate chemical balances; increasing ocean temperatures; and damaging many diverse ecosystems and the unique life present there. As educators, we arguably need to encourage our students to critically explore the intersection of people and ocean, learn from the things we discover through that exploration, and learn to adjust our lifestyles accordingly. We need to genuinely learn to live not just by the sea, but with the sea, for in the developed world we have not yet managed to do that.

Ko au te hone moana, ko te hone moana ko au.
I am the ocean swell, and the ocean swell is me.

Acknowledgements

I thank the students that I have been lucky enough to work with, for you have taught me much. I would also like to thank Hemi Hoskins for advice on Te Reo (Māori language).

Notes

1 In Aotearoa NZ, formal education comprises primary (years 1–8), secondary (years 9–13) and tertiary (university and polytechnic). I work at a polytechnic called Ara Institute of Canterbury (Ara), before 2016 known as Christchurch Polytechnic Institute of Technology (CPIT). CPIT is referenced in the following discussions.
2 Although this general trend has been observed in Aotearoa NZ, it might not be as visible in other countries. In the UK for example, the Royal Geographical Society (which is involved in working with the General Certificate of Secondary Education and A level syllabus and examinations) has embraced 'Changing Places' as a new strand of the curriculum.
3 Year 12 and 13 are the final two years of high school in Aotearoa NZ, similar to the UK system.
4 Completion of a journal that detailed their reflections on learning was a requirement of the course, and students have allowed their work to be used here.
5 Dave Rastovich is a professional surfer and environmentalist who has gained a significant profile through activism for the oceans, particularly for promoting the protection of dolphins and other marine mammals (for example, see Leitch, 2009).
6 For example, the promoting vegetarian diets (or reduced red meat diets) as a response to climate change, critical mass cycling to highlight the needs of urban cyclists, and the rejection of single use disposable coffee cups to minimise resource use.
7 The theory and place of action competence in the Aotearoa New Zealand curriculum is discussed in more detail by Chris Eames in Chapter 5).

References

Bolstad, R., Joyce, C. and Hipkins, R. (2015). *Environmental Education In New Zealand Schools: Research Update 2015*. Wellington: New Zealand Council for Education Research.

Capra, F. (1982). *The Turning Point: Science, Society and the Rising Culture*. London: Harper-Collins.

CPIT. (2013). Student movement to ban the bottle. [Online] www.cpit.ac.nz/news-and-events/items?a=178523 (Accessed 18 November 2015)

D'Arcy, P. (2006). *The People of the Sea: Environment, Identity, and History in Oceana*. Honolulu, HA: University of Hawai'I Press.

Della Porta, D. and Diani, M. (2006). *Social Movements: An Introduction* (2nd edition). Malden, MA: Blackwell Publishing.

Diamond, J. (2005). *Collapse: How Societies Choose to Fail or Succeed*. New York: Viking.

Eames, C., Law, B., Barker, M., Iles, H., McKenzie, J., Patterson, R., Williams, P., Wilson-Hill, F., Carroll, C., Chaytor, M., Mills, T., Rolleston, N. and Wright, A. (2006). *Investigating Teachers' Pedagogical Approaches in Environmental Education that Promote Students' Action Competence*. Wellington: NZCER.

Elliot, R. (2014). *Shark Man: One Kiwi Man's Mission to Save our Most Feared and Misunderstood Predator*. Auckland: Random House.

Fisher, A. and Ussher, B. (2014). A cautionary tale: What are the signs telling us? Curriculum versus standards reflected in schools' planning. *New Zealand Journal of Teacher's Work*, 11(2), 221–231.

Goodman, M. (2013). Ban the bottle. *New Zealand Surfing Magazine*, 158(89).

Hamilton, C. (2010). *Requiem for a Species: Why we Resist the Truth About Climate Change*. New York: Earthscan.

Irwin, D. (2015). Tightening the purse strings: A discussion about funding of EOTC in Aotearoa New Zealand. *Out and About* (31), 17–21.

Irwin, D. and Brasell, A. (2014). Serious learning and serious fun: Weaving achievement standards into outdoor education. *Out and About* (29), 21–25.

Jensen, B. B. and Schnack, K. (1997). The action competence approach in environmental education. *Environmental Education Research*, 3(2), 163–178.

Leitch, T. (2009). Dave Rastovich on ecology. *Adventure Magazine*, April/May, 14–23.

Meduna, V. (2015). *Towards a Warmer World: What Climate Change Will Mean For New Zealand's Future*. Wellington: BWB Texts.

Milbrath, L. W. (1989). *Envisioning a Sustainable Society: Learning Our Way Out*. Albany, NY: SUNY Press.

Parliamentary Commissioner for the Environment. (2015). *Preparing New Zealand for Rising Seas: Certainty and Incertainty*. Wellington: Parliamentary Commissioner for the Environment. [Online] www.pce.parliament.nz. (Accessed 3 March 2018).

Rahmstorf, S. and Richardson, K. (2009). *Our Threatened Oceans*. London: Haus Publishing Ltd.

Rankin, I. (2008). *Goodreads*. [Online] https://www.goodreads.com/search?q=ian+rankin&search%5Bsource%5D=goodreads&search_type=quotes&tab=quotes (Accessed 20 January 2008).

Rockström, J., Steffen, W., Noone, K., Persson, A., Chapin, F. S., Lambin, E., Lenton, T. M., Scheffer, M., Folke, C., Schellnhuber, H. J., et al. (2009). Planetary boundaries: Exploring the safe operating space for humanity. *Ecology and Society*, 14(2. Online] http://www.ecologyandsociety.org/vol14/iss2/art32 (Accessed 3 March 2018).

Rudd, X. (2012). *Follow the Sun. On Spirit Bird*. New York: Sony-ATV Music Publishing LLC.

UNESCO. (2004). *Millennium Ecosystem Assessment Report*. Paris: UNESCO.

Wearden, G. (2016) More plastic than fish in the sea by 2050, says Ellen MacArthur. *The Guardian*. [Online] https://www.theguardian.com/business/2016/jan/19/more-plastic-than-fish-in-the-sea-by-2050-warns-ellen-macarthur (Accessed 3 March 2018).

Whitehead, M. (2014). *Environmental Transformations: A Geography of the Anthropocene*. New York: Routledge.

Wright, R. (2004). *A Short History of Progress*. Toronto: House of Anansi Press Inc.

12 Bodies and technologies

Becoming a 'mermaid': myth, reality, embodiment, cyborgs, windsurfing and the sea

Barbara Humberstone

This chapter considers alternative definitions of the human condition: how it is shaped and mobilised through specific embodied sea-based practices and is realised socially and potentially 'politically' to give rise to greater environmental awareness and action. I take as a starting point for this discussion the entanglement of bodies and technology through the activity of windsurfing. My stories of being with the sea from a very early age are narrated elsewhere (see Humberstone, 2015), but here I begin by exploring the integrating nature of windsurfing, where I become 'at one' with the elements, the board and the sea. The board and I become embodied as 'mermaid'. I explore historical and mythical notions of human beings' relations with the sea, drawing on notions of the cyborg, part human/animal, part machine (Haraway, 2006 [1985]). There are vast numbers of engaging biographical-historical stories of the sea told largely from seafarers', generally male, perspectives. Many early seafarers told tales of seeing and hearing mermaids, which, as narrated in the Greek Iliad, lured mariners to their death. Cyborg myths, for example, or mermaid myths specifically, tell us of humans' continued interest in and longing for the ocean where humans become part animal, part fish and are thus able to inhabit the ocean.

These yearnings are carried across time, space and place, capturing the diverse maritime imaginations. A mermaid is a mythical creature that is part woman, part fish. Variations of the mermaid myth exist throughout the globe and have their own ideological and spiritual historical roots in the mists of time. Significantly, these folklores disrupt the boundaries between human and non-human and afford agency to these human/non-human sea creatures. These myths transcend space and time. Half-human/half-fish supernatural beings embody something of the energies of being and becoming one with the elements experienced when 'at/in/on/with' the sea. Paying attention to space–time re-imaginings (Foucault, 1986) and Haraway's (2006 [1985]) classic 'cyborg manifesto', I argue that as one windsurfs, surfs or sails one becomes sea-cyborg/mermaid or merman, connecting empathetically with elemental seascapes. In exploring the cyborg concept, the blurring of boundaries between human and non-human are explored. Such a conceptualisation is necessary.

I argue for the interconnection of all things. Through merging with the ocean we might recognise the significance of our place in the universe and become more sensitised to the interrelations of human and non-human nature. Experiencing Foucault's (1986) crisis heterotopia, the complexities of distorted space–time, enacted and experienced when the human and windsurf board become one with the sea, provides for these sensibilities.

Living in imaginative space–time of the sea

Inspired by Wright Mills' (1959) notion of the sociological imagination, I wish to re-imagine myths of the sea (such as mermaid myths) and current practices of being with, and in, the sea by drawing upon Soja's (1996) notion of third space and Foucault's (1986) closely related ideas of heterotopia, with De Certeau's (2007) notion of spatial practices of everyday life.[1] I draw on these theories because each – in their own way and together – provide for alternative, dynamic configurations of time and space that become reconfigured from conventional notions of either time or space. Soja's (1996) notion of third space, for example, has largely been associated with cultural urban spaces. Here I am concerned with the sea spaces in the light of these notions. Moreover, Foucault's concept of heterotopia highlights spaces that disrupt conventional order and normalising categories, thus destabilising taken-for-granted categories. Foucault's 'crisis heterotopia' has been envisaged as a space that may be regarded as a sacred or privileged place without geographical markers. I contend, though, that these theories have paid limited attention to corporeality or embodiment. Yet analyses drawing upon heterotopia and taking account of embodiment have been usefully utilised in exploring alternative physical cultures such as fell running and yoga practice spaces (Atkinson, 2010; Humberstone and Cutler-Riddick, 2015). Both texts serve to deconstruct dominant signs and images in such a way that ascribed identities become blurred and 'on occasions, [the] body–mind dualism was resisted and a reconfigured ontology of body–mind integration was rehearsed … The tension between inner self and outer body was largely dissolved and mind and body integrated' (Humberstone and Cutler-Riddick, 2015, 1222). Yet this work could be expanded further, by voyaging out to sea.

The sea, a dynamic and fluid space, may afford similar blurring of boundaries and reconfigurations. For example, space and time may merge as the windsurfer becomes part human/part equipment, illustrating permeable bodily configurations. Likewise, Anderson (2012) talks of the convergence of surfers, their boards and the motion of the waves in his study of surfers, identifying the significance of mobile place in analyses. In the context of windsurfing, the word 'windsurfer' can be used for both equipment (board and rig) and the person who windsurfs. In language, human and equipment are merged. The geographical location for windsurfing is generally along a shoreline where surfing can occur. Windsurfing, however, does not need surf waves for it to be practised, although many windsurfers choose surf locations. Wind, however, is essential.

'Learning to be in the body' through the spatial practices of everyday *sea*life (De Certeau, 2007), when windsurfing in and on the sea, affords a complex, dynamic relationship between the windsurfer's body, equipment (board and rig), sea and wind. Arguably, unlike any other form of sailing (except perhaps kite surfing) the body is part of the rig. A sailing boat can 'sail' without a connecting body. Large yachts and small dinghies can move with the wind in their sails, the helmsperson providing direction. Yet as Reason (2015) muses, even they can be replaced by an electronic/mechanical auto-helm. However, when windsurfing without the body linking the board to the sail, the gear will not operate and the windsurfer will not move purposefully through the water. Dant (1998) points out that sailing a windsurfer on the plane integrates the windsurfer's body as part of the rig (mast, sail and boom). To create movement of the board across the water requires both fine and gross body movements – interactions between the sailor and the board and rig. Holding the rig via the boom, the sailor's body (torso and hips), along with movement of the feet and toes, are integral to sailing the board. Steering is a process of balancing the sail to match the force of the wind. Through delicate movements of the body, leaning away from the sail (in effect a counterbalancing act) and applying pressure through the feet, control is exerted. This is somewhat akin to steering a skate-, snow- or surfboard but without the wind energy. Anderson (2012) likewise identifies the balance required to surf a wave.

Common to many water-based sports, the body interacts with equipment to move through water, creating sensations of speed and control. However, for the windsurfer, speed and control can only happen through the ways in which the body moves in relation to the rig and board, which are influenced by the state of the sea and wind direction and strength. On the plane, making tiny adjustments of the body in relation to the board and changing elements is split-second and intuitive, 'it must happen without conscious thought so that the equipment becomes like a prosthetic extension of the sailor's body' (Dant and Wheaton, 2007, 8). Thus, when 'planing' in windsurfing, according to this understanding, my body, my board and rig act as one; we become a cyborg of the sea's making. Likewise, Straughan (2012), in her autoethnographic account of sub-aqua diving and the relationship of touch to the sea sensoria, draws on Thrift to highlight the complex interconnections of human and non-human that become 'folded into the human world in all manner of active and inseparable ways, and most notably in the numerable interactions between things and bodies' (Thrift, 1999, 312, quoted in Straughan, 2012, 20).

Windsurfing is a temporal practice. It can take place over one to two hours or be short-lived if the conditions are not quite right for the windsurfer. Here I recall a very short event from my records:

> It is May but cold for the time of year. The tide is out in my usual sailing spot on the south coast of England so we decide to go to a different beach, which can take the WNW wind. We drive east on the motorway and then through narrow almost one-way lanes to the spot with free parking and neat rows of beach huts. Rigging up isn't the same, there is less space but more shelter from the wind. We carry the board and sail separately down to the water's edge. The tide has come up and the waves are breaking on the shore. There is one other windsurfer flying almost parallel with the shoreline. The

waves break as I try to get the board and the sail out through them. I wear a new thick wetsuit which stops the chill of the water reaching my skin. It feels comforting as the waves crash into me and pull and push the board away. I get onto the board and take off towards the yellow buoys. I feel the pull on my arms as the sail fills with wind and the board increases in speed. Not wanting to catapult, I don't slip into the lines but hold the boom with my hands and arms. Afraid to go out to sea too far, I gybe, and return back along the way I came; I jump off inside the tiny mini sand bar where the swell is inhibited. Back on the board to return to beach where I entered the water, I sail with the wind on the back quarter, the sail is further out and the board less stable. I roll down the waves feeling a sense of exhilaration as I am moving my body and gently tweaking the sail to surf, this is it surfing, down the small rolling waves until I reach the shore break and bring the board and sail ashore.

(Ethnographic Record, 2015)

Whilst Dant and Wheaton (2007) 'read' the practice of windsurfing through the lenses of Bourdieu and material and embodied (physical) capital – where significance is placed upon the cultural and economic (dis-)advantages gained through particular embodied practices – I 'read' windsurfing through 'seeking my senses', adopting the recent turn to the senses in physical culture (see Sparkes and Smith, 2012; Sparkes, 2017) to engage with sea space.

Embodied space–time

Through my windsurfing 'everyday' practices I am entering Soja's third space and Foucault's heterotopic space. My narrative above is brief, capturing only a few moments on/in the water, yet it still evokes bodily sensations and emotions: the warmth and protection that my wetsuit gives from the crashing cold sea; the ever-changing tides that provide different affect/effects of the sea, simultaneously creating fear and exhilaration as I momentarily ride the small wave back into the shore. Rarely do I feel the sensation of riding a wave but when this happens there is something of a singular sense of greater engagement with the sea than when the water is flat and the wind is the dominating force. Many surfing texts identify this oneness with the sea. Capp (2004, 11/86) a surf writer quoted in Anderson (2012) expresses this:

Whenever I tried to pin down what this 'at oneness' felt like, one particular moment in the surf always came to mind … I remember the water swelling beneath me and how I was perfectly in tune with its rhythm. I remember a surge of energy lifting me high above the hollowing water, the thickness of the shoulder, the glowing, desert-like appearance of the shore. Above all, I remember the instant at the top of the wave, just as I rose to my feet to 'take the drop', poised on the brink with the weight of the inrushing ocean behind me and the wave unfurling beneath me. The spool of my memories always froze at this last split second of clarity and separateness before the screaming descent where mind, body and wave – became one.

(Capp, 2004, 11/86, cited in Anderson, 2012, 581)

However fleeting, likewise for me the sensation is smooth and continuous. As the following account of windsurfing later that same year in my 'home' waters attests:

> I turn away from the wind slightly and manoeuvre effortlessly on the waves and then swing the sail through the gybe, balancing the board through the tips and troughs of the small waves. Beautiful feeling of flowing movement back towards the mini-spit, surfing along the little waves, moving the board at various angles to play on the wave … All my senses are keen … I'm balancing the sail in the wind with the board as it moves with the waves and I feel myself 'surf' along the small waves.
>
> (Humberstone, Fox and Brown, 2017, 90)

We perceive what sea space is through our senses. We see hear, smell, feel, taste, sense and intuit this wet space. As we move in sea space, time moves with us. A sense of balance is crucial to windsurfing and other physical practices in which equipment and body move on or through the sea. The body moves intuitively to maintain equilibrium. Balance is what is required to maintain our combined (board and body) movement in synchrony with the ocean's movements. The ocean's movements are a consequence not only of the elements such as wind but also the gravitational pull of the sun, the moon and the planets. As we subconsciously/intuitively 'balance' the board on the waves, we are engaging with the gravitational pulls of the cosmos. This occurs even when not 'surfing' but merely moving through the water. In all these situations, we are connected within the space–time continuum through each of our energetic senses. As Evers explains in his autobiographical account of being on/in the sea:

> By a lived body, I am referring to an active process, in which there is no unified biological body that is composed of permanent things and separate to the social. There is no definite boundary to determine where the body begins and external nature ends. A lived body recognises how there are occasions of experience that are not rooted in some unified ego in which I can claim a body as my own. The lived body never proceeds as some identical state as new elements and relationships intervene.
>
> (Evers, 2009, 904)

The body, then, becomes permeable, for:

> [b]eing in or on the sea attends to the whole body, not the (un-)conscious-ness in isolation but the whole of the corporeal body: mind, senses, their inter-relatedness and particular embodied relationship with the sea … we

become part of the fluid motion of the sea and are kinaesthetically and emotionally moved by and with it instantaneously and over time, long-term.

(Humberstone, 2015, 28–29)

The permeable body becomes part of the external environment, the senses align or interact with the cosmos to maintain equilibrium and enact movement. My kinaesthetic sensations on the sea speak to the flow and balance of my permeable energetic body with board and ever-changing fluidity of the sea, the wind in the sail and the gravitational pull of the cosmos. We experience the energies of the sea and the cosmos simultaneously, being in time–space. I enter into a *third space* in which there is an alternative configuration of time and space, a feeling of time standing still.

On occasions, different terms have been used by windsurfers, surfers and others to describe feelings of time standing still, of sharp awareness, of increased consciousness on experiencing singular 'magic' kinaesthetic occurrences. Some use the term 'flow', others 'oneness' or 'feeling connected with nature'. A windsurfer in Dant and Wheaton's discussion states: "'It's almost a spiritual thing [...] the feel good factor is so ... The simple physical feeling it gives you is great I think'" (2007, 11). Taylor's work (2007) further highlights the expressions of spirituality by surfers, oftentimes articulated as a religion. These enhanced sensations identified by surf and windsurf writers coincide with my own impressions and are reminiscent of Foucault's crisis heterotopia in which the sea is/becomes a sacred space without markers or boundaries and the windsurfer/sea-cyborg/'mermaid' being is dissolved into the space–time continuum of the cosmos.

Mermaids, myths and cyborgs

As noted, the windsurfer is linguistically *both* the equipment and the person who sails. Haraway's 1980s writings – drawing on notions of cyborg – challenge folk myths and dominant ideologies in their bolstering of oppressive relationships (such as patriarchal relations and those of dominion over nature). She challenges traditional dichotomised thinking which creates such distinctions as nature/culture, male/female and so forth. For Haraway the cyborg is a recognition/creation of the blurring of boundaries between the human and more-than-human worlds. Haraway described the cyborg creature as 'the figure born of the interface of automaton and autonomy' (1989, 139), which exists 'when two kinds of boundaries are simultaneously problematic: 1) that between animals (or other organisms) and humans, and 2) that between self-controlled, self-governing machines (automatons) and organisms, especially humans (models of autonomy)' (Haraway, 1989, 139).

Further, according to Haraway, 'the cyborg is a kind of disassembled and reassembled, postmodern and personal self' (2006 [1985], 130). Through this argument she calls for the deconstruction of composed binary opposites toward an understanding of all life, human and non-human, as equally valued. Through her vision of communication, she foretold in the 1980s the significant influence of world communication as we now know it in the World Wide Web. Whilst she couldn't foresee current events, her work, initially from a scientific perspective, talks to the significant changes in society and the profound effects and emotional *affect* of new forms of communication and the need for social and environmental justice.

She speaks to this principle through her cyborg manifesto. Through the notion of cyborgs she deconstructs 'origin' myths that she perceives as promoting forms of oppression. I suggest, however, that what she is deconstructing are 'Western'- interpreted forms of folklore that have already been reinterpreted/rewritten/ shaped through the process of Enlightenment thinking to work for the dominant ideologies of the time. For example, Hokowhitu (2007) heavily critiques the New Zealand film *Whale Rider*, adapted from the book of the same name by Ihimaera (1987), for its Westernised interpretation of Māori culture and the representations of the Whangara location. In contrast, Harmer (2003) challenges Haraway's notions of mythical creatures and myths as necessarily understood as oppressive in interpreting the fantasy story *Salt Fish Girl* (Lai, 2002), that draws upon the major Chinese origin myth and fantasies around human–animal-like creatures (mermaids) that she deems cyborg. This interpretation questions Haraway's assumption that myths are essentially 'deterministic doctrines that bind us to older ways of thinking' (Harmer, 2003, 1).

The sea cyborg, embodied as mermaid, and other mythical creatures predate modernity and modern science. Clearly these myths emerge from the mists of times. Half-human, half-animal fabled creatures emerged with the myths of creation. Various indigenous cultures retain their creation myths in which the world, human and non-human nature emerged out of darkness from nothing to become 'something'.[2] The mermaid myth, in particular, continued into the eighteenth century when voyaging seafarers claimed sightings of mermaids and mermen. Some of the then scientific community (see Radford, 2017) were fascinated with organisms that seemed to cross distinct categories and were part human and part fish. After a number of hoaxes, where the fraudsters tried to make money from claiming to possess a part-fish/part-human creature, scientists rejected the reality of mermaids and similar creatures. Nevertheless, at this time mermaid and mermen were given life in 'modern' fairy stories, most notably in the Danish myth written by Hans Christian Andersen about a little mermaid who wanted to leave her life in the sea and gain a human soul in order to be loved by a local prince. In so doing, she got legs by cutting out her tongue. In the tale, things don't work out well and she returns to an indeterminate life in the sea.

Dahlerup's (1991) five interpretations or deconstructions of this fairy tale draw upon various social and psychological frameworks to elicit the diverse meanings from the story, none of which look to eco-perspectives in their explorations, thus overlooking a wider, more connected interpretation of the mermaid myth. These analyses represent the social turn from scientific ideas to the exploration of popular culture (for example, Williams, 1958). The myth of mermaid creatures persists in popular culture. These became linked to dreams and fears and the focus on the position and place of male and female in society. Even today, 'mermaid' sightings are followed earnestly by enthusiasts and there is a small group who don a material fish tail and 'become' mermaids (Serk, 2017). We see the ways in which the mermaid (part human, part fish, a sea cyborg) has been consciously appropriated by modern humans and perhaps unconsciously through our engagement and convergences with the ocean as windsurfers or surfers.

Myths and reality

A Westernised take on the Māori traditional story of *Pānia of the Reef* in New Zealand gives a flavour of the mermaid image as pan-global,[3] and illustrates how this is a pre-Westernised mythical creature:

> Pānia was a sea maiden who swam ashore at sunset and returned to the sea before dawn. She would hide in a clump of flax beside a freshwater spring at the foot of Hukarere cliff in Napier. One day Karitoki, a chief in the area, was thirsty and came to the spring for a drink. He caught sight of Pānia, and took her home to be his wife. But every morning Pānia would return to the sea. After some time Pānia had a son, Moremore, who was without hair. The chief worried that he might lose his son and wife to the sea people. He consulted a tohunga who told him that if he placed cooked food on the mother and child while they slept, they would never return to the sea. He did this, but the ritual did not work and Pānia was turned into a rock, forever in the ocean. Moremore became a *taniwha* in the form of a shark. He lived in a cave in the sea, and his descendants used to frequent the Ahuriri harbour. He was a kaitiaki (guardian), patrolling the coastal waters and inner harbours while his people fished and gathered seafood.
>
> (Taniwha, n.d.)

In Māori culture the *Taniwha* is a mythical being and is seen on the one hand as a protector of the land–sea space, like Moremore in the story above, and on the other as a fierce and frightening creature.[4] One might identify particular *Taniwhas* as 'bio-cyborgs' since they may be part-human and part-animal/fish creatures, often having changed from human to fish/reptile form or vice versa. According to Strang (2014), *Taniwhas* (and other indigenous mythical creatures, such as Trolls in Iceland) have been successful in some situations in protecting parts of the land, rivers and coastline from exploitation. Strang (2014) asserts that in New Zealand, Aotearoa:

> A Māori bioethics of partnership with nonhuman is well expressed in descriptions of the *Taniwhas* that are believed to inhabit key water places. A *Taniwha* is: … a living being whose spirit remains present at the spring and at certain places along the stream … The *Taniwha* is a generative 'life essence' of people and places … encapsulating ideas about shared substance and social constructions between people and places. The well-being of the

Taniwha is connected to the well-being of the people ... and harm to the *Taniwha* or its home is believed to have impact on [their] health and well-being.

(Strang and Busse, 2009, 4)

Arguably, when windsurfing, surfing or sailing, one engages with the cosmos through balancing the gravitational fields; as time stands still, on occasions when windsurfing one becomes a 'sea cyborg' or 'mermaid/man'. Something of the 'life essence' of an imagined *Taniwha* creature potentially comes into play through the interconnected energies of the elements that become embodied, thus distilling and embedding forms of 'kinetic empathy' (Thrift, 2008, 237).

Through 'everyday' (De Certeau, 2007) sea practices the windsurfer experiences the embodied energies of the cosmos in heterotopic time–space, which Taylor (1999), when referring to surfing, talks of as 'communion with nature'. Myself (see Humberstone, Fox and Brown, 2017) and other sea practitioners (highlighted in, among others, Taylor, 2007; Anderson, 2012) experience these energies through their senses and often express sentiments of this as what might be interpreted as sacred or numinous experiences which – on occasion through their bodily practices – have been articulated as 'spiritual'. For example, I have expressed this in my autoethnographic writing:

> I feel the water rushing past my feet and legs. The wind in my hair. I sense the wind shifts in strength and direction and move my body in anticipation to the wind and the waves. I feel the power of the wind and the ability of my body to work with the wind and the waves. The delight and sensation when surfing down a small wave with the sail beautifully balanced by the wind. Seeing the sea birds and the fish jump delight further. The smell of salt and mud. The small seal that made its home on the tiny pebble spit. These are some of the beauties of windsurfing in this liminal space.
>
> (Humberstone, 2015, 35)

Thrift (1999) draws attention to the notion of 'kinetic empathy', a concept or sense that Thorpe and Rinehart (2010) develop in relation to nature-based sport practitioners, to highlight social and environmental action undertaken by some water-based activity enthusiasts. I suggest that this '[k]inetic empathy is crystallised at the embodiment, senses, emotions and practice-in-nature nexus experienced amongst communities of practitioners across the globe and transmitted and actualised in the twenty-first century mobile space and time dimensions' (Humberstone, 2011, 506).

Such a speculative concept as kinetic empathy – emerging from sociological spatial imaginations – has caused us to ponder upon the reality of actions amongst communities of windsurfers, surfers and other sea people who promote environmental action at local level and more globally. In a sense, there is a 'tuning into' *cosmic* or *universal* energies. Observed through conventional Western objectivist and positivistic understandings of everyday life and practices, this is perhaps unthinkable. Yet it is certainly a possibility if it is framed in Eastern philosophy, where heterotopic spaces are reality and we can become 'one with the Universe' in physical practice.[5]

Political reality

At the present time of increasing resource scarcity and powerful multinational acquisition, Strang points to the complexities and real inequalities of land and water 'ownership' (2014, 129). She states that 'majority populations are being disenfranchised as governments sell publicly owned water to private elites, simultaneously dissolving their own capabilities to uphold the common good' (Strang, 2014, 129). Māori and other indigenous traditional values,[6] which she points out may well have been displaced in many cultures, are at odds with the neoliberal view of water as merely economic resource. Consequently, she argues that in order for the common good and indigenous values – here interpreted as social justice and environmental sustainability – to prevail, it is necessary for some degree of control to be gained over water. Protests in 2010 against coastal mining along the New Zealand, Aotearoa coastline brought together Māori and non-Māori activists and these actions were, initially at least, partially successful. Many of the surfing and sea-based communities were involved in these actions. However, looking ahead toward a greater 'imaginative' caring global governance, the current political hedonism in continuing ascendance is deeply disturbing.

Concluding remarks

This is a speculative piece in which I have drawn together some of my thoughts, embodied practices and various texts related to the sea, water, windsurfing, mythical creatures and their interconnections and interpretations in the context of space–time cosmic thinking to highlight the permeability of human and non-human relations and the significance of our engagements with the ocean. This has been to engage the reader to think imaginatively and holistically about the connections of bodies and technologies, of myths and realities, and of social and environmental justice and the ways in which we are all merged with the planet and the cosmos. In the process of writing this it becomes much clearer to me how powerful the seas and water flowing into them are. It also strikes me how far industrialised cultures have become removed from these watery roots and from their connections with the energies of nature, water, the sea and the cosmos.

Re-imagining and re-storying *myths* and recognising the role of embodied experiences may provide for some protection of the human and more-than-human worlds that reside in, on, under and by the sea. I would suggest that to understand the ocean, it is necessary to dissolve the binaries that separate humans and the non-human; to think of space and time in situ; and to recognise that we are intimately connected and merged with the ocean. Surely this is one of the greatest challenges that people of the sea face, as we attempt to live with the sea in a rapidly changing world.

Notes

1 The stimulus for this time–space perspective emerged from Maier's (2013) paper concerned with time–space and social geography in emergent Christianity.
2 This has some symbolic links with understandings of the 'big bang' and other current scientific notions concerning the creation of our universe. But is arguably unlike the notions of 'creation' of relatively recent Western Creationists who claim scientific validity for their view of the literal interpretation of the Bible.
3 The global reach of the mythical mermaid is also evidenced in Japan, where body-boarders like to express and identify themselves as mermaids. Here the symbolism is feminised but also expressed physical skill and activity (Mizuno, 2018).
4 In UK, a very local 'myth of the croc' has been storied to frighten young children from going too close to a pond in a local outdoor education centre where adults think they might fall in and drown (see Humberstone and Stan, 2012).
5 It is worth noting developments in quantum mechanics such as Minkowski's space–time continuum which is predicted as a fourth dimension. This idea links space and time as singular dimension and predicts mathematically time moving backward as well as forward and the possibility of 'wormholes' – corridors through which other parallel universes can be found (see Penrose, 2005).
6 Strang and Busse (2009, 8) refer also to Canadian and Australian indigenous peoples who have particular claims to land and water. She summarises all their common factors: individual and collective identity is based on shared substantial connections to specific ancestral beings or deities, places and the waters of those places; collective rights of water ownership and use are based on this connection; there is no place, in these cosmological schemes, for alienation of people from their land and resources; human well-being and environmental well-being are connected, and disturbance of the environment, such as impediment of proper water flows, is understood to have social, ecological and spiritual impact.

References

Anderson, J. (2012). Relational places: The surfed wave as assemblage and convergence. *Environment and Planning D: Society and Space*, 30(4), 570–587.

Atkinson, M. (2010). Entering scapeland: Yoga, fell and post-sport physical cultures. *Sport in Society*, 13(7), 1249–1267.

Dahlerup, P. (1991). Splash. *Scandinavian Studies*, 63(2), 141–163.

Dant, T. (1998). Playing with things: Objects and subjects in windsurfing. *Journal of Material Culture*, 3(1), 77–95.

Dant, T. and Wheaton, B. (2007). Windsurfing: An extreme form of material and embodied interaction? *Anthropology Today*, 23(6), 8–12.

De Certeau, M. (2007). *The Practice of Everyday Life*. Berkeley: University of California Press.

Evers, C. (2009). 'The point': Surfing, geography and sensual life of men and masculinity on the Gold Coast, Australia. *Social and Cultural Geography*, 10(8), 893–908.

Foucault, M. (1986). Of other spaces. *Diacritics*, 16(1), 22–27.

Haraway, D. (1989). *Primate Visions: Gender, Race, and Nature in the World of Modern Science*. New York: Routledge.

Haraway, D. (2006 [1985]). Manifesto for cyborgs: Science, technology, and socialist feminism in the late twentieth century. In J. Weiss, J. Nolan, J. Hunsinger and P. Trifonas (Eds.), *The International Handbook of Virtual Learning Environments*. Netherlands: Springer Press, 117–158.

Harmer, E. C. (2003). Myths of origin and myths of the future in Larissa Lai's *Salt Fish Girl*, Forum: 'Origins and Originality' [Online] http://forum/11C.ed.uk (Accessed 14 January 2016)

Hokowhitu, B. (2007). Understanding Whangara: *Whale Rider* as simulacrum. *New Zealand Journal Media Studies*, 10(2), 53–70.

Humberstone, B. (2011). Embodiment and social and environmental action in nature-based sport: Spiritual Spaces, Special Issue – Leisure and the politics of the environment. *Journal of Leisure Studies*, 30(4), 495–512.

Humberstone, B. (2015). Embodied narratives: Being with the sea. In M. Brown and B. Humberstone (Eds.), *Seascapes: Shaped by the Sea: Embodied Narratives and Fluid Geographies*. Farnham: Ashgate, 27–40.

Humberstone, B. and Cutler-Riddick, C. (2015). Older women, embodiment and yoga practice. *Ageing and Society*, 35(6), 1221–1241.

Humberstone, B., Fox, K. and Brown, M. (2017). Sensing our way through ocean sailing, windsurfing and kayaking: Tales of emplaced sensual kinaesthesia. In A. C. Sparkes (Ed.), *Seeking the Senses in Physical Cultures: Sensual Scholarship in Action*. Oxford: Routledge, 81–100.

Humberstone, B. and Stan, I. (2012). Nature in outdoor learning-authenticity or performativity well-being, nature and outdoor pedagogies project. *Journal of Adventure Education and Outdoor Learning*, 12(3), 183–198.

Ihimaera, W. (1987). *Whale Rider*. London: Penguin.

Lai, L. (2002). *Salt Fish Girl*. Toronto: Thomas Allen Publication.

Maier, H. O. (2013). Soja's Thirdspace, Foucault's heterotopia and de Certeau's practice: Time–space and social geography in emergent Christianity. *Historical Social Research*, 38(3), 76–92.

Mizuno, E. (2018). Multiple marginalization: Representation and experience of bodyboarding in Japan'. In lisahunter (Ed.) *Surfing, Sex, Genders and Sexualities*. Abingdon: Routledge, 71–90.

Penrose, R. (2005). Minkowskian geometry. Chapter 18 in *Road to Reality: A Complete Guide to the Laws of the Universe*. New York: Alfred A. Knopf

Radford, B. (2017). Mermaids & mermen: Facts & legends. [Online] https://www.livescience.com/39882-mermaid.html (Accessed 17 September 2017).

Reason, P. (2015). Sailing with Gregory Bateson. In M. Brown and B. Humberstone (Eds.), *Seascapes: Shaped by the Sea: Embodied Narratives and Fluid Geographies*. Farnham: Ashgate, 101–108.

Serk, L. (2017). Mermaids gather to compete in UK competition. [Online] http://www.bbc.co.uk/news/uk-england-40675250 (Accessed 24 August 2017).

Soja, E. (1996). *Third Space: Journeys to Los Angeles and Other Real-and-Imagined Places*. Oxford: Blackwell.

Sparkes, A. C. Ed.(2017). *Seeking the Senses in Physical Culture. Sensuous Scholarship in Action*. London: Routledge.

Sparkes, A. C. and Smith, B. (2012). Embodied research methodologies and seeking the senses in sport and physical culture: A fleshing out of problems and possibilities. *Qualitative Research on Sport and Physical Culture Research in the Sociology of Sport*, 6, 167–190.

Strang, V. (2014). The Taniwha and the Crown: Defending water rights in Aotearoa/New Zealand. *WIREs Water*, 1, 121–131.

Strang, V. and Busse, M. (2009). *Taniwha Springs-Indigenous Rights and Interests in water: Comparative and international perspectives, Advisory report for Nga Puna Wai o Tokotoru*. Auckland: New Zealand (Confidential Legal Report).

Straughan, E. R. (2012). Touched by water: The body in scuba diving. *Emotions, Space and Society*, 5, 19–26.

Taniwha – creature of many forms. (n.d.). New Zealand Myths and Legends. [Online] https://www.pinterest.co.kr/wakaNINE/new-zealand-maori-myths-legends/(Accessed 13 January 2016).

Taylor, B. (2007). Surfing into spirituality and a new, aquatic nature religion. *Journal of the American Academy of Religion*, 75(4), 923–951.

Thorpe, H. and Rinehart, R. (2010). Alternative sport and affect: Non-representational theory examined. *Sport in Society*, 13(7/8), 1268–1291.

Thrift, N. (1999). Steps to an ecology of place. In J. Allen and D. Massey (Eds.), *Human Geography Today*. Cambridge: Polity Press, 295–321.

Thrift, N. (2008). *Non-representational Theory: Space, Politics, Affect*. London: Routledge.

Williams, R. (1958). *Culture and Society*. New York: Columbia University Press.

Wright Mills, C. (1959). *The Sociological Imagination*. Oxford: Oxford University Press.

13 Past and presents

Making connections with the sea: a matter of a personal and professional *Heimat*

Mark Leather

The sea is my *Heimat*; my homeland, my place, my sense of belonging in, by or on the sea. It is a place that is familiar, comfortable and comforting. This chapter is an autobiographical account of my connection with the sea, seascapes and ultimately my/our maritime cultural heritage. I suggest it is useful to consider this autobiographical approach not just as a record of my life, 'but as life itself' (Moss, 2001, 19). Moss (2001) considers how autobiography is a process, not only of recording (in the sense of documenting, orienting and analysing) but also of *becoming*, in the sense of lives, subjectivities and identities. While this critical self-reflection is not all-inclusive, or the only way of knowing, 'it can be a helpful and workable approach in gaining insight into one's life as well as into the contexts within which one exists' (Moss, 2001, 20).

In presenting my autobiographical reflections in this chapter, I share my story with the possibility that the reader may gain insight, and that some potential common life experiences resonate and provide a foundation for mutual themes of understanding about connections to the sea. These interpretive themes can be considered as 'meaning-metaphor-milieu' (Buttimer, 2001, 34); where *meaning* refers to my professional activity, *metaphor* to cognitive style and *milieu* to the environmental features of my childhood and formative years, where, in all three of these themes, there is scope for mutual understanding. In this sense, then, autobiography is not only a data source or research approach, 'it can also assist in critique and theory building' (Moss, 2001, 8). In making sense of these autobiographical themes, I am inspired by the approaches of Richard Shusterman (see 2000, 2008, 2012), the pragmatist philosopher, and the existential phenomenologist Maurice Merleau-Ponty (2012 [1945]), who provide some insight into the aesthetic and the embodied experiences discussed here.

In this chapter, I consider how my past and present personal relations to the sea as my *Heimat* have directly influenced my professional practice as a university lecturer. My construction of reality is what Bruner describes as 'hermeneutic composability' (1991, 8); that is, the telling of my story and its comprehension *as* a story depends upon the human capacity of the reader to process knowledge in an interpretive way. Autobiography is popular in feminist approaches to scholarship as a category of studying and practice where individuals are compelled to display self-knowledge through the creation and presentation of stories about the self (Cosslett, Lury and Summerfield, 2000). Autobiography is to do with recovering a past, 'and depends on the deployment of an often shifting, partial and contested set of personal or collective memories" (Cosslett, Lury and Summerfield, 2000, 4). Questions of validity inevitably arise with the use of autobiography due to the subjective nature of memory recall and the possibility of embellishment. Van Manen (1997) robustly defends this approach, describing how it may include fictitious scenarios, emotions or moods within a life story, as this allows us to be imaginatively or emotionally involved. In this sense, it is my intention that the reader may find some resonance, insight and understanding in the *meaning, metaphor* and *milieu* presented in this chapter. This connection to seascapes starts with my first recollections of encounters with the sea.

Personal connections

My connection to the sea starts with my annual family seaside holidays to Broadstairs on the Kent coast in south-east England for the first 11 years of my life. As I write, I recall that such holidays were, in many respects, very different over 40 years ago. I grew up in Enfield (of Lee Enfield[1] and Royal Enfield[2] fame), a part of suburban north London, in what seemed to me like an endless urban sprawl of houses, roads, shops and factories. This urban landscape was interspersed with recreational sports fields and the occasional town park, a green space often accompanied by ornamental ponds in which I could try to sail my toy boat. My parents, both Londoners, were aspirant middle class[3] people who espoused hard work and the importance of a good education. They were 'older parents' for that time: my father was 44 and mother 38 when I arrived to join my brother, who was three years my senior. I grew up with the stories of their adult experiences of the Second World War as ever present in the narratives of my childhood. As well as the cultural values of shared suffering for the common good, looking out for the neighbours, taking care of family and being thrifty with money and possessions, this soundtrack contained their personal narratives – my mother as schoolgirl during the blitz; my father as a volunteer in the Royal Navy. One year – I must have been ten years old – my father started to tell me his stories about serving in the Royal Navy as an enlisted man during the Second World War. These were the best history lessons. I gained a sense of connection with boats, ships and being at sea, although I had no direct experience of it. I grew up with my father mixing his naval slang ('keep the paint off the deck son') with the remnants of his childhood cockney rhyming slang ('use your loaf son'). And I suggest that these early experiences connected me to my British naval heritage. I have argued elsewhere (Leather and Nicholls, 2016) how I believe that our cultural heritage is firmly rooted in the mindset of an island nation.

My early family years were full of routine and tradition and I think it would be fair to say my parents were not naturally adventurous. Each weekday my father would commute to work in central London, leaving early and returning in time to say goodnight. My mother was the homemaker. Washing day, cleaning schedules and the family menu were pretty much identical each week. Weekends were filled with music lessons, shopping for groceries and jobs around the house and garden. Summer weekends were spent 'over the club' – the tennis club with grass courts where my parents met. They played tennis whilst we played around the open spaces with the other children. The highlight of the summer was the one-week seaside holiday to Broadstairs on the English Channel – and this was identical for the first 11 years of my life.

Our annual family holiday was eagerly awaited, and the familiarity of our holiday routine added to the excitement. It was always for one week; and to make the most of our time away we would be woken in the dark, having packed the bags and the car the previous night, and drive the two-hour journey to have a picnic breakfast on the beach. There was the traditional game of 'who can see the sea?' This added a sense of excitement at our imminent arrival. We always stayed in the same guest house (my parents never tried camping[4]), we went to the same beach every day (where the swimming was safe), a beach hut was always rented to base ourselves, and dinner in the guest house was served every evening at 6.30 in the evening. Despite these routine and the repeated experiences, or perhaps because of them, I remember these holidays as truly happy times as a family, times which created a regular, repeated and routine connection with the sea.

Figure 13.1 illustrates this connection. This image illustrates my early connection to the sea. I was five years old and this is one of my earliest memories. This is not the stereotypical image of British seaside holidays, with blue skies, sandcastles and ice cream. Even if we had used a colour film in the camera (black and white was cheaper), this was a day for long trousers and a coat. It was a cool day, and the drizzle and grey skies reflected my mood. It was our last day and we were due to go home, and I remember feeling a great sense of loss. I believe I remember the emotions (sadness at leaving, disappointment over the rain and no beach on my last day) and the smells and tastes (the sea, the damp, the saltiness) – although of course it may have been the repeated telling of the story when viewing this photograph in the family album that locates this image firmly in my memory and my connection to the sea.

Making sense of intimate connections

Aesthetic experience, cognitive understanding of place and my own cultural history have shaped my connection with the sea. Growing up in middle class suburban London, as discussed above, my connection to the natural world was perhaps impeded by the urban landscape. I learnt to swim in a local pool, and enjoyed playing in the highly managed town parks and sports grounds at school and during my leisure time. I had no knowledge and little contact with nature, aside from my mother's birdfeeders, and no regular access to open water. The only stimulus I can find was my annual family seaside holiday and the somewhat romanticised[5] stories of my father, as outlined above. I have now lived in Devon for most of my adult life, and close to the coast for all of that time. I cannot afford a home with a sea or estuary view, but I am only a short walk to my

Figure 13.1 Seaside holidays Broadstairs, Kent, England. Image owned by the author.

nearest beach or local harbour. I have lived on a boat and taught sailing, learned to kayak, surf, paddleboard, and play in, on and by the water as often as I can for work and for pleasure. My favourite activity is sailing, not to race and compete but to journey, to explore and arrive in a harbour via a form of transport that defines and connects me to what I see as my maritime cultural heritage – one that is founded upon seafaring.

My connection with the sea clearly starts from my early and repeated childhood experiences, rather than any historical appreciation. These early childhood experiences have been described by Chawla (1998, 373) as *significant life experiences*, and she concludes from the research of others that 'childhood experience in the outdoors is the single most important factor in developing personal concern for the environment'. Perhaps it was these early significant outdoor experiences by the sea that started my personal connection, interest and concern with seascapes. I suggest that the reader looks back to their own early experiences and engages in a similar autobiographical journey to see how the past influences their present engagement with the sea, and the many emotional connections that can come with it. Perhaps these connections are strong because the sea has a special place in our memories of happy times shared with loved ones. I am also curious as to whether other sensibilities emerge from these emotional ties – such as a concern for the planet or, more specifically, the state of plastic in our oceans (for example, see Decker, 2014; Thompson, 2017).

My cumulative memories of these seaside holidays are brought alive the moment I smell the sea. There is a distinctive, familiar and reassuring smell of saltwater, rock-pools and seaweed that presses my olfactory button and releases emotions and memories deep within me. When these smells are accompanied by the sight of a large open expanse of sea, the endless horizon and the taste of salt in the air, it is these aesthetic experiences that are strongest, and it is this that helps me understand this deep-felt, emotive connection with the sea. These sensory experiences of the sea are similar to the geographer Edensor's (2010) discussion of sensory rhythm in urban wildscapes, including how the smells affect us. Edensor (2010) explores how our spatial connections are built through the senses, and how our visual consumption of landscapes is expanded into a more holistic consumption of sensory rhythms, textures, sounds, smells and tastes, and I argue here that the rhythmicity he explores is equally applicable to our sensory experiences of seascapes. Moreover, the work of Steinberg and Peters (2015), where they make the assertion that critical perspectives can be gained by turning to the ocean, and in particular the ocean's very geophysical character, is also pertinent. They are clear that the ocean is not simply a liquid space; it is solid (ice) and air (mist); it generates winds, which transport smells, and these may emote the 'maritime' miles inland. A sea that can be smelled travels with the air. This sea can infiltrate adjacent, connected spaces through the wind, which floats it to the shore in altered form. This more-than-wet ontology helps explore my way of being with the sea.

The work of Richard Shusterman, the pragmatist philosopher who developed the concept of *Somaesthetics* (see Shusterman, 2000, 2008, 2012) is also influential in making sense of my connections with the sea. In *Performing Live*, Shusterman describes how 'we cannot get away from the experienced body with its feelings and stimulations, its pleasures, pains and emotions ... all affect is somatically grounded' (Shusterman, 2000, 152–153). In doing so, Shusterman acknowledges the influence upon his thinking of William James, who over a century earlier argued that, 'a purely disembodied human emotion is a nonentity. If we try to abstract from any strong emotion all the feelings of its bodily symptoms, we find we have nothing left behind' (James, 1890, 173–174). There is also Merleau-Ponty (2012) with his central notion of embodiment, who asserts that we are not just a free-floating consciousness in the world, but an embodied being where the body is central to consciousness, knowledge, understanding and *intentionality*.[6] Additionally, Tim Ingold provides insight into the body and our body–world interactions (Ingold, 2011). He argues that rather than thinking of the inhabited world as composed of mutually exclusive hemispheres of sky and earth separated by the ground, we need to attend to the fluxes of wind and weather. To feel the air and walk on the ground is not to make external tactile content without surroundings but to mingle with them. As he writes, 'in this mingling, as we live and breathe, the wind light and moisture of the sky and bind with the substances of the earth in the continual forging of all way through the tangle of lifelines the compromise the land' (Ingold, 2011, 115). For me, this 'mingling' is a useful way of considering the complete, holistic experience of time spent outdoors, on the land, under the sky, in the weather and by, on or in the sea.

Certainly, a range of emotions are engendered in me through this embodied experience of 'mingling' with the sea. The smell of sea air; the taste of salt on my lips; the feel of the fine, wet sand under my feet and between my toes; the sense of power when a wave, bigger than expected, turned my world upside down and rolled me up the beach. Likewise, the 'bracing' nature of the cold English Channel for regular swims (long before the mass availability of wetsuits) and the burning sensation of salt water accidentally swallowed or breathed up the nose serve as good examples of embodiment, and these all helped connect me to the sea from a young age. Shusterman (2008) asks how we might combine critical body mindfulness – in other words an awareness of this sensate embodiment – with the demands for smooth spontaneity of action. Rather than some general self-awareness, he suggests that body consciousness and mindfulness enrich the present experience and enhance the continuity of experience, thereby creating an appetite for further encounters and opportunities to extend the consciousness we have of our body. It was by endorsing the need for reflection, whilst firmly embedding a role for mindfulness in relation to body consciousness, which enabled Shusterman to develop his concept of *somaesthetics* as a way of capturing this integrated process.

The connection between a somaesthetic experience, mindfulness and nature is an interesting and contemporary topic. Hyland (2009) considers mindfulness to be part of the so-called 'therapeutic turn' in education. He suggests that that there are some educationally justifiable goals underpinning this therapeutic turn. He argues that the concept of mindfulness can be an immensely powerful and valuable notion, which is integrally connected with the centrally transformative and developmental nature of learning and educational activity at all levels.

My experiences of the sea and seascapes are embodied, aesthetic, cognitively conscious and mindful. The level of conscious cognition has increased over the years since my childhood holidays. I now use a range of sensing activities in my practice of outdoor education with young adults at university and try to balance the embodied experiences with those of conscious self-reflection in order to facilitate opportunities of 'nature connectedness' for my students, so that they may be attentive to nature, themselves and the sea by observing and noticing their senses (see the second section of this chapter).

For me personally, however, the sea has the ability to provide 'felt immediacy', more so than a hike in the mountains. Perhaps this is the power of the sea and the oceans, or 'water worlds' (see Anderson and Peters, 2014). For example, the splash of cold salt water in my face as I paddle my kayak out through the surf is 'felt immediacy' and I feel at home. Anderson highlights how kayakers enjoy lifestyles that are tied to moving water connected through the emotions felt through immersing with it and being mobile on it: 'kayaking on the sea gives humans the opportunity to be other than who they are on land, and gives a new way of looking at the world as a consequence' (Anderson, 2014, 114). Similar is the all-embracing cold sensation (in English waters) as I jump into the sea for a swim from a yacht. Water has an inimitable facility to bring me into the present – the 'immediate' here and now. It also, of course, unfailingly demands respect and my careful attention.

Whilst acknowledging the importance of felt immediacy, this begins to challenge any assumption that my activity by/in the sea focuses upon absolute control of my body. Instead this idea is reversed to suggest there is a need to become bodily attentive, conscious and mindful in relation to my environmental transactions. We can never control or master the sea as its agency is beyond our manipulation, and as such we have to work with it (Peters, 2012). For example, when subjected to a powerful wave breaking over me and rolling me up the beach, I definitely have a lack of command over my body and the element of water. The challenge posed here is related to a shift in ideas from that of *commanding* the body to the development of an ability to *listen* to it in everyday situations. This moves the idea of command of the body in action to a heightened consciousness of it and introduces the concept of bodily sensibility. Although listening to our bodies in the throes of a demanding activity may be possible, there are limits to explicit consciousness during these experiences. The following example illustrates this point. Whilst kayak surfing, having just learned to 'eskimo roll', I underestimated the size of the surf and overestimated my own level of competence in a boat. This resulted in me paddling up a wave, flipping over backwards and – in attempting to roll up the force of the water – caused me to fall out of my boat and be tumbled ashore in the white messy broken wave. I was certainly not master of the sea. Clearly, I was not commanding my body, yet retained a body awareness and sense of the power of moving water. The physical challenge of the sea and my connection with it is deeply rooted in my embodied experiences of it. Feeling the power of the wind with all the sails filled whilst on a high-performance dinghy, cruising yacht or gaff-rigged fishing trawler allows me to 'magically' move across the sea's surface. The challenge of reading the currents and tides around the coastline are part of an engagement and immersion 'in being' with the sea. The sea commands from me a great deal of respect as well as a sense of awe and wonder. The risk of drowning is ever present, as is that of hypothermia; the conduction of heat away from the body when immersed is far greater than in air and the shock of cold water on the human body has an 'immediacy' – a felt reality. Yet at the same time, in my experience, there is a wonderfully alluring and at times mystical quality to water. There are a number of somaesthetic stimuli: the visual picture of inky-smooth darkness, or of clear azure seas, or the white caps of waves as they break on the beach; the smell of the sea; the feel of seaspray on my face; the taste of salt in the air; the feeling of near weightlessness that allows me to turn somersaults when swimming; and the exhilaration of jumping in from rocks whilst coasteering[7] and surging back to the surface.

Finally, the somatic self is always situated in the physical world. Therefore, somatic awareness cannot really be of the self alone. This supports Merleau-

Ponty's perspective of nature that reveals the indivisibility of human and nature and the inseparableness of mind and body (see also Chapter 12, this volume). In his classic *Phenomenology of Perception* Merleau-Ponty (2012 [1945]) explores experience to reveal the primordial relationship of body-subject and the world. Sailing in stormy waters or walking along a beach, making footprints in the sand in the pouring rain, obliges us to sense the ground or water beneath us, to be sensitive to the forces of nature acting on our bodies. Ingold (2010) discusses how, as footprints are made in soft ground, rather than stamped on a hard surface, their temporality is due to the dynamics of their formation and these are a function of the weather and of reactions across the interface between the earth and air. As he notes, 'breathing with every step they take wayfarers walk at once in the air and on the ground. This walking is itself a process of thinking and knowing. Thus, knowledge is formed along paths of movement in the weather world' (Ingold, 2010, 121). Our minds and bodies are inextricably connected to and interwoven with the fabric of the world so that mind/body/world are closely intertwined and mutually influencing.

Forming these felt experiences into consciousness comes via thorough reflection, something which is intrinsic to the act of expressing the essence of those experiences. For my professional practice as a teacher, the past – social dimensions, the aesthetic experiences and cognitive reflection upon them – are interwoven into the fabric of my teaching in the present. Such reflections may be the sharing of stories of a particularly thrilling sail, or extremely rough channel crossing, with friends or students around the dinner table onboard ship or by a campfire on the beach. It may also come from the sharing of images taken while actively engaged in time at sea. I find myself using images to help convey these experiences for me and others by sharing them – more recently sharing images widely utilising various forms of social media (Quay, 2013). In the next section of the chapter I move from my personal to professional *Heimat*, drawing connections between them.

Professional connections

In this section, I explore how the activity of sailing provides a means with which to connect students to maritime cultural heritage by exploring and studying the places where the land meets the sea: the towns, harbours and quays where we go sailing.[8] Ryan's (2016) detailed exposition of this land and sea interface considers coastal explorations of landscape, their representation and associated spatial experiences. She discusses the awareness of the relationship forged between self and surroundings. Teaching the skills of dinghy sailing in the fishing port of Brixham, south Devon, my colleague and I encouraged our students (via an assessed piece of coursework) to explore and research the rich history of the town, to help connect them with seafaring heritage and more specifically to understand the concepts of 'place' as well as specifically their own connection to Brixham (see Cresswell, 2015; Leather and Nicholls, 2016). For example, one student reflected upon the once large fleet of sailing trawlers. We had visited one of the remaining trawlers, *Provident*, shown in Figure 13.2, lying alongside the harbour wall. They wrote:

Figure 13.2 Brixham Trawlers. Image with permission. Provident.

'[w]hen I was on the water at Brixham or at the sailing centre, I would look out at the harbour and imagine the 300 vessels [once] overpowering the area with their [ochre] red sails' (in Leather and Nicholls, 2016, 454).

The experience clearly had an influence on students' sense of place, cultural heritage and connection with the sea. These two extracts from their work explain their embodied, immediate, conscious reflections:

> I think the impact of having a unique and challenging experience in [a] place with a sense of immense historical and cultural presence, especially if it is relevant to your experience, can be a lot more sensory and you discover more about yourself and those others in your group. Sometimes it feels real education can be lost in the speed of modern life. ...
>
> The smell of the salty harbour of which you feel you can taste, the sound of the wind whistling through the sailing boats masts, the smell of fresh fish and chips on the quay and the wonderful views of colourful layered houses. I would end every session looking out into the harbour and appreciating just how beautiful Brixham was as a place and how glad I was to experience it.
>
> (Leather and Nicholls, 2016, 454, 461)

The students' writing expressed the importance of planning and facilitating their learning to encompass this embodied intellectual, social, personal, cultural and

Figure 13.3 A wet and windy day in Plymouth Sound. Photo by the author.

historical self-understanding. I continue to make the most of every opportunity to engage them – as shown in Figure 13.3 on a trip to the 200-year-old Plymouth Breakwater, part of the naval defences for the military port.[9]

Making sense of work-based connections: romanticised seascapes

Finally, in this chapter I consider the tension that exists, when teaching about the maritime pasts and cultural heritage, between a (potentially) romanticised view of seafaring and seascapes and the historical, social, political issues of power, class and gender. In his cultural history of the sea, Mack (2013) discusses how a style of national history writing emerged that might be described as romantic. It reduces the attention to the sea as a globalised transnational space to one that is primarily focused on the rise of national navies and their colonial, imperialist ambitions, which – because they are instruments of statehood – tend to be expressed in terms of admirals and captains performing good deeds in defence of the nation and the sea battles in which they and their fleets have participated. There is an engrained triumphalism evident in some of this literature. As Mack notes, '[t]he extensive documentation of the transgressive history of pirates give some sense of alternative aspects to the recounting of maritime history ... it is too often cast in terms which, in the same vein, can appear overly romantic' (Mack, 2013, 20).

My romanticised view, especially of my childhood seaside holidays, is a factor of which I am aware when I am teaching my university undergraduates. Legg defines nostalgia as 'the longing for a home that no longer exists – or never existed' (Legg, 2004, 100). It is the longing for happy childhood days spent with parents who are no longer alive. The photographs are limited, and yet they manage to intensify the stories when re-told. They were happy days that first connected me to the sea, and certainly my father's stories painted an exciting and romanticised picture (in my mind) of his life in the Royal Navy as a young man, drawing a connection to my maritime past, in the present. In present times, I am mindful that my educational and recreational excursions on the sea are 'playing' at seafaring (Woodyer, 2012). I am not dependent on wind, tides and good seamanship to make a living or to travel around. I engage with the sea out of choice rather than necessity, teaching students about *place-based outdoor educa-tion*. As Legg also highlights, 'space and place are examined not just as weak metaphors but as formative factors in thinking about the presentness of the past' (Legg, 2004, 106).

British history is full of stories about the sea for adventures, exploration, trading and the conquering of people, lands and possessions. However, the telling of these stories privileges the wealthy and has little to do with the common seaman, as Lemisch states, 'in that *mischianza*[10] of mystique and elitism, "seaman" has meant Sir Francis Drake, not Jack Tar; the focus has been on trade, exploration, the great navigators, but rarely the men who sailed [the] ships' (Lemisch, 1968, 372). However, Adkins and Adkins' (2009) comprehensive account of life in Nelson's navy paints a bleak picture of life on board for the sailor known as Jack Tar. The conditions below decks were very different from those of the privileged officer class who had to be guarded by armed marines whilst they ate their dinner in the officers' mess in case of mutiny. Issues of class and power were exacerbated by the presence of the 'pressed men' found below decks, pressed men being taken only from the working classes not the merchant or upper classes. The National Museum of the Royal Navy (2014) describes impressment in matter-of-fact terms. Impressment was a long-standing authority from the state for the recruitment to military service at sea. The Impress Service, known as the press gang, was employed to seize men for employment at sea in British seaports. Impressment was used as far back as Elizabethan times, when this form of recruitment became law. Officially, no foreigner could be impressed although they were able to volunteer. If, however, the foreigner married a British woman, or had worked on a British merchant ship for two years, their protection was lost and they could be impressed. However, these limits were often ignored and the impressment of Americans into the British navy became one of the causes of the American War of 1812 (National Museum, 2014). This forcible conscription remained law past the point of the abolition of slavery in the UK, and was regarded similarly by the families and men affected by impressment. As Rediker has noted, '[t]he seaman's belief was that impressment violated his rights as guaranteed by the Magna Carta' and this remained 'an ever-present concern to Jack Tar' (Rediker, 1989, 34).

Sir Francis Drake, an iconic figure of British naval history, and Admiral of the Royal Navy, was also a privateer. A privateer was a pirate who by commission from the government was authorised to seize or destroy a merchant vessel of another nation. The privateer was used as a cheap means of weakening the enemy by frequenting shipping routes. Privateers were like private contractors; they received a Letter of Marque from their nation's Admiralty, which granted them permission to raid enemy ships and keep a percentage of the spoils – so long as they paid a part of that prize back to the government. In theory, no privateer with a Letter of Marque from government could be charged with piracy, since it was recognised by international law. These 'great men of history' (after Thomas Carlyle, 1840) were popularised by historians in the nineteenth century and arguably given folk-hero status, which I suggest directly affects our collective cultural consciousness about Great Britain, great men and naval traditions to the present day. The reality of pirates in history as well as the present day, is distorted by the stories associated with them in popular culture. Pirates are often understood in terms of fantasy, romance and the thrill of dangers now safely confined to the past. The stories of piratical deeds in films and novels are seen as real, or unreal, as the histories of pirates' lives, where fact and fiction merge in tales of Long John Silver and Blackbeard (Ogborn, 2008). Romanticised ideals of a pirate's life are everywhere in literature and film, none so more than the highly successful Johnny Depp as Captain Jack Sparrow in the Pirates of the Caribbean franchise. These images and portrayals clearly engage with our affection for 'loveable' rogues, connect us through the transmission of stories and perhaps help connect us to a culturally held seascape. It is important to remember that piracy today remains a real violent threat to shipping in parts of the world's oceans. It is also necessary to situate history's pirates within the discussions of imperial politics, New World settlement, global trade and maritime society and culture. While this may be entertaining, it perpetuates these cultural stereotypes. I consider the problems of this past, romanticised view of the sea in more detail below.

My father was raised in the time of the British Empire and I am certain he held the belief that British imperialism was a force for good – and being British was great! His stories of the Second World War (including the little ships of the Dunkirk evacuation and the Atlantic convoys) continued this cultural transmission. However, exploring my cultural heritage leads me to consider how I 'must penetrate beyond the romantic saga' (Sager, 1989, 74). As a child, I was taken to the tourist attraction the *Cutty Sark*, one of the last tea clippers to be built and one of the fastest of the time, coming at the end of a long period of design development which stopped as sailing ships gave way to steel ships and steam power. At that time, I watched *The Onedin Line* on television with my family. This was a BBC drama that was broadcast from 1971 to 1980 (Evans, 2006). The series dealt with the rise of a shipping line, and depicted the lives of the Onedin family, providing an insight into the lifestyle and customs at the time, not only at sea, but also ashore. The series also illustrated some of the changes in business and shipping, from wooden sailing ships to steel steamships. It also revealed the role that ships played in such matters as international politics, uprisings and the slave trade (Evans, 2006). My memories are of beautiful looking sailing ships and an air of adventure and excitement. Clearly a TV drama is not a historical representation, for as (Sager, 1989,72) states, 'there is nothing romantic about the nineteenth century coastal schooner or timber ship'. Sager's main aim in *Seafaring Labour* (1989, 4–10) was to bring the seaman into the mainstream of social and economic history, to situate 'labour at sea in the transition to industrial capitalism', a process that transcended 'sea and land and national boundaries'. To do so he challenged the romantic view of seafaring and those who write lovingly of ships and the sea. Sager's work tells the stories of sailors' lived experiences as part of an industrial process. Machinery was central to sailors' working lives and relationships, 'for they knew it with all five senses, with hands and muscles' (Sager, 1989, 5). This greater understanding of mine, the substance beneath the popular cultural representations that I acquired in childhood, has led me to question the essence of my affection for the sea.

As I reflect back on the stories and images of my youth I am conscious that seafaring was portrayed as a mostly conspicuously male-dominated occupation. It was only in 1993 that women were officially allowed to go to sea on board Royal Navy vessels (BBC, 2011). In Nelson's time, Admiralty Regulations stated that women were not allowed to be taken to sea. Whatever the rulebook said, it is clear that women did travel aboard Nelson's ships. From wives and respectable passengers to prostitutes and mistresses, women were very much part of shipboard life in the eighteenth and nineteenth centuries (BBC, 2011). The work of Creighton and Norling (1996) explores the relationship of gender and seafaring specifically in the Anglo-American age of sail. They show how popular fascination with seafaring and the sailors' rigorous, male-only life led to models of gender behaviour based on 'iron men' aboard ship and 'stoic women' ashore. Creighton and Norling (1996) also provide new material that defies conventional views. They investigate topics such as the role of the captain's wife onboard ship and women in the American whaling industry.

Women were not confined to secondary roles. Anne Bonny and Mary Read are arguably the two most famous women pirates (Ogborn, 2008). They were captured in 1720 and put on trial in Jamaica. The two women certainly did not seek to hide themselves on board, putting themselves forward for the most dangerous duties. Witnesses at the trial testified that they both very profligate, cursing and swearing freely, and willing to do anything on board. Others said that they only wore men's clothes to fight. Women wanting to live as pirates usually disguised themselves as men, similar to the 'transvestite heroines' (De Erauso, Stepto and Stepto, 1996) who dressed as men to serve on the crews of sailing ships. Bonny and Read appeared to hold their own on a pirate ship, and were respected by the rest of the crew. According to Ogborn, when they were captured it was noted that they were fighting at close quarters and 'these two with just one other kept the deck'; the rest of the crew were too scared and were hiding below decks (Ogborn, 2008, 190). Bonny and Read understood the value all common seaman, and pirates in particular, placed on courage, and they used it as a weapon against the men who might have exercised power over them

(Ogborn, 2008). Such stories not only tell the diverse stories of life at sea in detail, but in doing so have expanded my own understanding of the connections of gender, class, colonisation and race at sea and on land. These stories present a picture that is very different to my romantic view of seafaring generated from popular culture, and from my own past.

The work of Sager (1989), Rediker (1989), Creighton and Norling (1996), Ogborn (2008) and Adkins and Adkins (2009) has raised my awareness of the realities of life at sea, and challenged the romanticised views that were conveyed through my childhood and popular culture. While I have only had limited space here to discuss the issues of gender and the sea, as above, it is indicative of how much more there is to cover in my journey through seafaring and seascapes. I am now more mindful in my teaching about the potential for portraying a romantic view and perception of the sea, seafaring and great adventures and so perpetuate these stereotypes. I now take the opportunity to tell the stories and engage in dialogue about the masculine and class-oriented maritime representations and these issues of social justice. This can be in a lecture or during informal chat, conversation and dialogue whilst sailing a boat in front of Plymouth Hoe. Importantly, I continue to seek out and explore the diverse views of seascapes that my students bring with them from their own experiences.

Epilogue

In this chapter I have briefly shared stories from my past and present of my personal and professional connections to seascapes, from my first encounters as a child at the beach to my current practice educating university students studying adventure education and outdoor learning. I have made sense of this in part by considering the embodied, aesthetic experiences and the cognitive connections to the 'places' where I live, work and play, as well as interrogating the role of my British cultural seascape heritage. I am aware that for my journey into this history I have but dipped my toe in the waters. I intend to continue this exploration for my own connections with seascapes. This chapter is presented here to share my journey and passion for the sea, and to continue to understand my own story. My intention is to encourage others, not just my students but the readers of this chapter, to understand and engage with the aesthetics and narratives that the sea has to offer, through their own embodied culturally informed experiences, their own pasts as well as the present in order to make sense of their own visceral, emotional and cognitive connections to the sea.

Notes

1 The Lee–Enfield bolt-action rifle was the main firearm used by the military forces of the British Empire and Commonwealth during the first half of the twentieth century. It was the British Army's standard issue from 1895 until 1957 (see Skennerton, 2007, 90).
2 Royal Enfield motorcycles: the legacy of weapons manufacture is reflected in the logo, a cannon and a motto, 'Made Like A Gun' – 'The Bullet', a single cylinder bike that was also exported around the British Empire and is still manufactured in India

3 In my childhood, a stereotypical middle-class family may consist of two parents, married, living in semi-detached or detached house in suburbia, with a car in the garage and a well-maintained garden. They would likely have two children, be certain about the importance of a 'good education' and have aspirations that their children would attend university and then enter one of the professions.

4 Although they were highly encouraging and supportive of my time in the scouts and the frequent camps we attended.

5 My father did not romanticise the war or his time spent in the Royal Navy. I did as a young man, and have since become aware of this.

6 Following the work of Edmund Husserl, Merleau-Ponty's intentionality held that normal perception involves a consciousness of place tied to one's capacities for exploratory and goal-directed movement, which is indeterminate relative to attempts to express or characterise it in terms of 'objective' representations – though it makes such an objective conception of the world possible (Stanford Encyclopedia of Philosophy, 2002) According to Merleau-Ponty, then, intentionality is not mental representation at all, but skilful bodily responsiveness and spontaneity in direct engagement with the world.

7 Coasteering is a physical activity that encompasses movement along the intertidal zone of a rocky coastline on foot or by swimming. A defining factor of coasteering is the opportunity provided by the marine geology for moving in the 'impact zone' where water, waves, rocks, gullies, caves etc. come together to provide a very high-energy environment.

8 My teaching has been influenced by works such as *A Pedagogy of Place* (Wattchow and Brown, 2011).

9 Plymouth Breakwater is a 1,560-metre stone breakwater protecting Plymouth Sound and the anchorages near Plymouth. Around 4 million tons of rock was used in its construction in 1812 at the cost of £1.5 million.

10 Italian similar to mischiarsi – mixture, blend.

References

Adkins, R., and Adkins, L. (2009). *Jack Tar: Life in Nelson's Navy*. London: Abacus.

Anderson, J. (2014). What I talk about when I talk about kayaking. In J. Anderson and K. Peters (Eds.), *Water Worlds: Human Geographies of the Ocean*. Farnham: Ashgate Publishing, pp. 103–118.

Anderson, J. and Peters, K. (Eds.). (2014) *Water Worlds: Human Geographies of the Ocean*. Farnham, Surrey: Ashgate Publishing.

BBC. (2011). *History: Women in Nelson's Navy* [Online] http://www.bbc.co.uk/history/ british/empire_seapower/women_nelson_navy_01.shtml (Accessed 31 May 2015).

Bruner, J. (1991). The narrative construction of reality. *Critical Inquiry*, 18(1), 1–21.

Buttimer, A. (2001). Home-reach-journey. In P. Moss (Ed.), *Placing Autobiography in Geography*. Syracuse: Syracuse University Press, pp. 22–40.

Carlyle, T. (1840). *On Heroes, Hero-Worship, and the Heroic in History*. [Online] http:// www.gutenberg.org/ebooks/1091 (Accessed 20 November 2016).

Chawla, L. (1998). Significant life experiences revisited: A review of research on sources of environmental sensitivity. *Journal of Environmental Education*, 29(3), 11–21.

Cosslett, T., Lury, C., and Summerfield, P. (Eds.). (2000). *Feminism and Autobiography: Texts, Theories, Methods*. London: Routledge.

Creighton, M. S., and Norling, L.(Eds.). (1996). *Iron Men, Wooden Women: Gender and Seafaring in the Atlantic World, 1700–1920*. Baltimore: Johns Hopkins University Press.

Cresswell, T. (2015). *Place: An introduction*. Chichester: John Wiley & Sons.

De Erauso, C., Stepto, M., and Stepto, G. (1996). *Lieutenant Nun: Memoir of a Basque Transvestite in the New World*. Boston, MA: Beacon Press.

Decker, J. (2014). *Gyre: The Plastic Ocean*. London: Booth-Clibborn Editions.

Edensor, T. (Ed). (2010). *Geographies of Rhythm: Nature, Place, Mobilities and Bodies*. Farnham: Ashgate Publishing.

Evans, J. (2006). *The Penguin TV Companion*. London: Penguin.

Hyland, T. (2009). Mindfulness and the therapeutic function of education. *Journal of Philosophy of Education*, 43(1), 119–131.

Ingold, T. (2010). Footprints through the weather-world: Walking, breathing, knowing. *Journal of the Royal Anthropological Institute*, 16(1), 121–139.

Ingold, T. (2011). *Being Alive: Essays on Movement, Knowledge and Description*. London: Routledge.

James, W. (1983 [1890]). *Principles of Psychology*. Cambridge, MA: Harvard University Press.

Leather, M., and Nicholls, F. (2016). More than activities: Using a 'sense of place' to enrich student experience in adventure sport. *Sport, Education and Society*, 21(3), 443–464.

Legg, S. (2004). Memory and nostalgia. *Cultural Geographies*, 11(1), 99–107.

Mack, J. (2013). *The Sea: A Cultural History*. London: Reaktion Books.

Merleau-Ponty, M. (2012 [1945]). *Phenomenology of Perception*. Trans. D. A. Landes. London: Routledge.

Moss, P. (Ed). (2001). *Placing Autobiography in Geography*. Syracuse, NY: Syracuse University Press.

National Museum (2014) Impressment [Online] http://www.royalnavalmuseum.org/sites/default/files/Impressment.pdf (Accessed June 2014)

Ogborn, M. (2008). *Global Lives: Britain and the World, 1550–1800*. Cambridge: Cambridge University Press.

Peters, K. (2012). Manipulating material hydro-worlds: Rethinking human and more-than-human relationality through offshore radio piracy. *Environment and Planning A*, 44(5), 1241–1254.

Quay, J. (2013). More than relations between self, others and nature: Outdoor education and aesthetic experience. *Journal of Adventure Education & Outdoor Learning*, 13(2), 142–157.

Rediker, M. (1989). *Between the Devil and the Deep Blue Sea: Merchant Seamen, Pirates and the Anglo-American Maritime World, 1700–1750*. Cambridge: Cambridge University Press.

Ryan, A. (2016). *Where Land Meets Sea: Coastal Explorations of Landscape, Representation and Spatial Experience*. London: Routledge.

Sager, E. W. (1989). *Seafaring Labour: The Merchant Marine of Atlantic Canada, 1820–1914*. Montreal: McGill-Queen's University Press.

Shusterman, R. (2000). *Performing Live: Aesthetic Alternative for the Ends of Art*. Ithaca, NY: Cornell University Press.

Shusterman, R. (2008). *Body Consciousness: A Philosophy of Mindfulness and Somaesthetics*. Cambridge: Cambridge University Press.

Shusterman, R. (2012). *Thinking Through the Body. Essays in Somaesthetics*. Cambridge: Cambridge University Press.

Skennerton, I. D. (2007). *The Lee-Enfield: A Century of Lee-Metford & Lee-Enfield Rifles & Carbines*. Labrador, QLD, Australia: Ian Skennerton Publishing.

Stanford Encyclopedia of Philosophy. (2002). Consciousness and intentionality. [Online] http://plato.stanford.edu/entries/consciousness-intentionality/ (Accessed 20 November 2015).

Steinberg, P., and Peters, K. (2015). Wet ontologies, fluid spaces: Giving depth to volume through oceanic thinking. *Environment and Planning D: Society and Space*, 33(2), 247–264.

Thompson, R. (2017). Environment: A journey on plastic seas. *Nature*, 547(7663), 278–279.

Van Manen, M. (1997). *Researching Lived Experiences: Human Science for an Action Sensitive Pedagogy* (2nd Edition). London, Ontario, Canada: The Althouse Press.

Wattchow, B., and Brown, M. (2011). *A Pedagogy of Place: Outdoor Education for a Changing World*. Clayton, VIC: Monash University Publishing.

Woodyer, T. (2012). Ludic geographies: Not merely child's play. *Geography Compass*, 6(6), 313–326.

14 Rituals and performance

Crossing the line: all at sea with King Neptune mid-Pacific

Robin Kearns

For as long as I can recall, the sea's depth has daunted me. Yes, I have long loved the immersion of swimming in the sea. But this love has only prevailed in the knowledge that I can touch sand or rock below. Since early childhood, the open ocean has been a 'between-ness', a watery and daunting highway connecting earthly outcrops and landmasses. My earliest memories are of a wide expanse: days that stretched out and became weeks aboard the passenger liner *Rangitane* at the age of four. Departing from London docks, we first crossed the Atlantic, then the Pacific, arriving in New Zealand, a new land that tentatively became home. Although memories are far from clear, I know that while at sea in 1963, the depth and seeming infinity of the ocean scared me; scarred me even. Where did the ocean begin and end? I felt out of my depth.

In the words of Symes (2012), 'journeys are narrative-like, with beginnings, middles, ends and sequels'. In this chapter, I explore two aspects of this ocean voyage from England to New Zealand: the middle (being at sea) and the sequel (living with an anxiety about deep water). I interrogate both my own memories and photographs of the voyage left by my late father. In so doing I build on the idea that autoethnography can be more than analysis of one's own written texts. Rather, it can include scrutiny of remnant documentation concerning the cultural connection between self and others (Chang, 2007). In this case, as a child-migrant, I was part of a family engaged in a life-changing transition from a familiar cultural context to another yet unknown one, a place of 'yet-to-be'. To that extent, life at sea was aboard a liminal space in which anything could – and, at times, did – happen.

To craft an account of this time, and its downstream effects, by relying only on fragments of pre-school memories would be, at best, an impressionistic enterprise. Hence, my selective recourse to photographs taken by my father have allowed me to 're-view' myself in a setting I have largely forgotten. This approach allows me to follow Butz and Besio in seeking to 'dissolve to some extent the boundary between authors and objects of representation, as authors become part of what they are studying' (2009, 1660). This point is also made by Muncey (2005, 55), who writes of the way artefacts such as photographs can act as stimuli, conjuring up feelings and thoughts.

My chapter focuses in particular on one significant event on the *Rangitane*'s voyage to Auckland: the mid-Pacific staging of the ritual of King Neptune's

Court. I draw on literature from the field of emotional geographies to argue that this occasion of maritime 'camp' and pantomime was, for me, a 'de-naturalising' experience unsettling memory and ontological security. For others, however, it can have different possibilities, such as ritually asserting dominance over the oceanic unknown, or a making of memories through creatively engaging in the carnivalesque at a place and time far from familiar shores. The broader goal of the chapter is to contribute to evolving understandings of blue spaces that extend from a focus on oceans to water-based leisure geographies (for example, Anderson and Peters, 2014; Merchant, 2011; Olive, McCuaig and Phillips, 2015; Spence, 2014) adding to recent creative approaches to, and representations of, what it is to live with the sea (for example, Brown and Humberstone, 2015; Schellhorn and Perkins, 2004; Stocker and Kennedy, 2011).

Fear and the sea

> *Anchor me*
> *In the middle of your deep blue sea, anchor me*
> *Anchor me, anchor me*
>   (Don McGlashan, 1994)

The sea has long generated feelings of both attraction and anxiety. It can be both literally and figuratively a source of buoyancy (Foley, 2015) as well as a threat to well-being through storms, rips and conditions too challenging for swimming. Historically, it has also been a gendered space. Feminised mermaid figures have mythically embodied fish-like features and exuded charm and seduction, whereas masculinised figures like pirates or King Neptune have embodied more fully human forms and symbolised dominance and mastery of the depths. Accounts of men's fear rarely enter the waters of what otherwise might be a recreational or work space (though see Chapter 8 in this volume); rather, writers such as Joseph Conrad have tended to associate constructions of masculinity with mastery of the fundamentally fluid space of the sea. Accounts of seafarers have shown how exemplary constructions of masculinity on shore and at home help them to endure fearfully harsh workplace conditions at sea (McKay, 2007).

Overarching discourses of masculinity ensure men rarely admit publicly to what they fear, least of all in academic print (Connell, 2005). Indeed, as Phoenix and Frosh describe, there is invariably a 'façade of fearlessness and bravado' (2001, 27) which is employed, especially by younger men, to assert their authority over females and male 'others'. Fears are usually discussed in the aggregate (for instance, 'men's fears') or uttered in private and, even in autobiographical accounts that acknowledge emotion, fear has been largely kept in the closet (Kearns, 2015).

For geographers, an interest in linking places and emotions such as fear has grown from the humanistic scholarship in the 1970s and 1980s and the psychoanalytic 'turn' of the 1990s (Davidson and Milligan, 2004). In their challenge to take emotions seriously, Anderson and Smith (2001) emphasised concern for lived experience and feelings through a focus on doing and performing, rather than simply representing emotions. As an outgrowth of both the 'emotional turn' and wider post-structural developments, geographers have embraced non-representational theory as a way of 'doing and writing human geography that aim(s) to engage with the taking-place of everyday life' (Anderson and Harrison, 2010, 1). A wide range of work now emphasises the emotional aspects of personal and social life (Pile, 2010), including engagement with the sea through practices such as sailing (Peters and Brown, 2017) and surfing (Anderson, 2014). Notably, however, most of this discussion of emotions and the sea points to positive valences and there is a gap in literature acknowledging experiences of fear, anxiety and trepidation with regard to the sea.

I am not alone in 'going public' with fear, however. Over the last two decades, geographers and sociologists have examined dimensions of fear relating to threats such as sexual violence (Pain, 1997) and wider geopolitical oppression (Pain and Smith, 2008). But personal accounts have been less common. What exactly is fear? At root it is an unpleasant feeling that may lead to behavioural changes (such as retreat from a situation) brought on by awareness of danger. To the extent that fear is the anticipation rather than experience of danger, it is a state of 'not yet'; it occupies a time and place of anticipating an occurrence that could or might happen, but equally might not (Pain, 2000). Fear lies within that gap of the possible or the possibly not and it is the unpredictability of outcome that produces fear and that retains hold over us.

This chapter's confession of intermittent anxiety in living with the sea, and its admission that depths rattle me, parallels Tolich (2012), who writes of overcoming a fear of flying after doing an Outward Bound course in New Zealand's Marlborough Sounds. With candour, he writes that 'I cannot understand the source of that fear, nor can I even recreate it. Something happened at Anakiwa that freed me from my fears' (Tolich, 2012,12). Similarly, I can identify sentinel times and places when fear has either been transcended or has insinuated itself back into my being.

As Tolich (2010, 13) notes, 'writing about the self in isolation is a virtually impossible task'. The idea for this chapter emerged in the context of an at-sea community of 15 (see Introduction). Two colleagues (the editors of this volume) encouraged me to write on the theme when, in a preamble to discussing a paper, I spontaneously mentioned my aversion to deep water. For 14 months I stalled drafting this chapter. First I was too busy. Then it was too hard. And too personal. Following the time at sea, the idea seemed to have returned to the depths of my subconsciousness. Once back on dry land, like a mollusc, the tendrils of my anxiety retreated until I just had to write before the tide of opportunity ebbed. These layered feelings of procrastination and reluctance to commit personal experience to a professional context reveals, perhaps, a second level of fear: that reluctance of the academic self to speak from the heart as well as the head.

This narrative account applies an autoethnographic perspective to memory, ambivalence and the sea. It departs from the usual applications of autoethnography that generally entail the interrogation of texts collected by the researcher for either personal (for example, a journal) or explicitly research purposes (such as field notes) (Butz and Besio, 2009). Do there need to be texts to analyse? Can there be an autoethnography of memory; a re-living of a time in which the only legacy is a few photographs and fragmented memories? The answer is surely 'yes' if, as Wall (2008) maintains, autoethnography is not only method but also a theoretical perspective that acknowledges the embodied place of the researcher and values social-scientifically pursuing personal questions. The second 'yes' is when there are discernible reverberations of events from the past that ripple out into the present casting subsequent times and places into sharp, if not anxious, relief.

In the case of my voyage with my parents on the *Rangitane* at the age of four, I have no texts on which to draw; only a handful of photographs and my own distant memories from five and a half decades ago. Clearly 'there are risks associated with the partiality of self-focused research' (Symes, 2012, 56), but in this case there is no imperative to speak for, or on behalf of, others. Rather this account is an attempt to give voice to an ambivalence for the sea that upwells from biography and a version of which may flow through the memories and encounters of others.

Seaborne mobilities

> *Along that wilderness of glass –*
> *No swellings tell that winds may be*
> *Upon some far-off happier sea –*
> *No heavings hint that winds have been*
> *On seas less hideously serene.*
> (Edgar Allan Poe, 1831)

Being aboard ship is an experience of 'in-between geographies, abeyant spaces where the normal parameters of time and space are suspended' (Symes, 2012, 59). Usual routines of home, school or work are temporarily suspended as passengers' days are shaped by the set mealtimes and on-board leisure activities. Paradoxically, as in the case of train travel (Officer and Kearns, 2017), the vessel moves and is a 'vehicle for knowing' (Hasty and Peters, 2012), taking its passengers with it. Yet those on board experience *immobility* given that most of the usual opportunities for movement are limited (Anim-Addo, Hasty and Peters, 2014; Symes, 2012). Indeed, a ship is a contained and potentially immured space (Weaver, 2005) such that one is 'held' for the duration of the voyage. Life is slowed while afloat, and decelerated (Vannini, 2013). Yet, as Symes (2012) points out, far from being a timeless space, a ship's roster of watches, mealtimes and crossing of time zones, as well as the need to keep to schedule, means that ships can be regarded as *time machines*. Temporal awareness on ships, if anything, is more pronounced not less when, for instance, time zones are crossed (Borovnik, 2004).

In terms of its own time line, the New Zealand Steamship Company's *Rangitane* was the second vessel of that name, the first having been sunk off East Cape in World War II. The newer *Rangitane* plied the London–Curacao–Panama--Papeete–Auckland route between 1949 and 1968. It was 21,800 tons in weight, 609 feet long, and had capacity for 460 passengers. Among them in 1963 were Michael Kearns, a 36-year-old English veterinary surgeon, his 32-year-old Scottish wife, Heather, and me – their four-year-old son. Farewells at the London docks had been bittersweet; a great adventure beckoned, but the passage was one-way and siblings and parents did not know when they would return. As was the custom, family could come aboard for a taste of what was to come before the horns blew and ropes loosened.

Crossing the line

The sea, in fact, is that state of barbaric vagueness and disorder out of which civilisation has emerged and into which, unless saved by the effort of gods and men, it is always liable to relapse.

(W. H. Auden, 1950)

It was a time before speed. A time I barely remember. A time long before the slow food and slow cities movements (Honoré, 2004). Slow was simply the way of getting places. There are few events I recall from that journey: the smell of oil and industry in Curacao; cowering in our cabins with a raging storm at sea (Figure 14.1); lowered flags for Kennedy's assassination; a downpour sending deckchairs afloat as we passed through the Panama Canal; and the lush tropical colours and smells of Tahiti. After a few weeks at sea, I recall a buzz of anticipation about crossing the equator. The days were hot and still (Figure 14.2).

My most memorable and somewhat traumatic recollection of that voyage was the ritual that occurred aboard the *Rangitane* at that time (Figure 14.3). As Richardson (1977) recounts, this ceremony, known as the 'Order of Neptune', is deeply anchored in maritime history, having been performed by sailors on ships in the Western world for over four centuries. The ritual consists of the initiation of the 'landlubbers' or 'polliwogs' (those who have never crossed the equator) by the 'seafarers' or 'shellbacks' (those who have previously crossed the line). It involves an ordeal in which novices must present themselves to King Neptune (a gaudily dressed up senior crew member), who has purportedly come aboard from the inky equatorial depths. Novices are subjected to a ritual death and rebirth through enduring various humiliations such as being pelted with food, or in the case of my recollection on the *Rangitane*, being injected by the ship's doctor with tomato ketchup and raspberry juice. This set of performances is undertaken in order to gain their new status of seafarer. A long-held tradition on military or commercial vessels, it has more recently been deemed to constitute hazing and is forbidden on some ships. Toned-down versions persist for the sake of entertainment on some contemporary cruise ships.

Figure 14.1 Storm at sea, aboard the *Rangitane*, 1963. Image owned by the author.

Figure 14.2 Calm day at sea with my mother Heather, aboard the *Rangitane*, 1963. Image owned by the author.

Figure 14.3 King Neptune holds court, aboard the *Rangitane*, 1963. Image owned by the author.

Two possible origins of the ritual have been proposed. First, it has been interpreted as marking the movement from 'landlubber' to 'seafarer' status. This transition is seen to be important so that seasoned sailors could be assured that their new shipmates were capable of handling the trials of being at sea. The second interpretation is that the ritual is a means of placating natural and fearsome forces at work in the sea.

To a four-year-old, witnessing crew dressed up as creatures of the deep, daubed in paint and shouting incomprehensibly had neither interpretive effect. Rather, at an age of firm belief in the tooth fairy and Father Christmas, this was a vision to let loose imagination of krakens waking and a sense of foreboding about all that lies in the depths. Further, the outsized syringes used to inject mock blood left me with a particular legacy of aversion to needles and having blood taken (Andrews, 2011).[1]

This ritual has deeply historical origins. Richardson (1977) reports that the earliest account of an equatorial crossing performance involved the French vessel *Parmentier*, which crossed the line in 1529 and whose crew offered prayers and tossed coins overboard. The ceremony soon took on more dramatic elements. By the late sixteenth century, there was a visit from 'Father Neptune', who demanded a tithe in silver from those who had not yet crossed the equator. If this bribe was not paid, the newcomer would be shaved and dunked into the sea from the yardarm or drenched with a barrel of sea water. As the ceremony has been maintained it has also evolved. For instance, the ship's captain has often been given the part of King Neptune and senior officers the roles of Davy Jones and Aphrodite within his court. This mock court is often preceded by a beauty contest involving men dressing up as women, with the occasion of reaching the equator being not only the crossing of an imaginary line but also a rare opportunity for cross-dressing, and diluting otherwise firmly gendered boundaries, at sea. Physical hardship, in keeping with the spirit of the initiation, is tolerated, and each Pollywog is expected to endure an initiation rite in order to become a Shellback. Appearing before the King, 'truth serum' (hot sauce and aftershave) and whole uncooked eggs are described by Richardson (1977) as being put in the mouths of novices. He goes on to report the use of grease, mustard and chilli sauce with the dunking of participants in tubs of water and a prevailing sense of humiliation through a requirement of Pollywogs to crawl on hands and knees in a stylised ritual death and rebirth. What were once equator-crossing ceremonies conducted as a serious initiation rite have, according to Richardson, morphed into King Neptune's courts that are staged for passengers' entertainment on civilian ocean liners.

Like a child at a horror movie, my own crossing-the-line experience left me more anxious than amused. Where had these creatures of the deep come from? Would they return? When? Was all deep water a space of dark possibilities of humiliation? If I fell into such depths, would I ever return?

On not taking the plunge: bathophobia

I was much too far out all my life; and not waving but drowning.
(Stevie Smith, 'Not Waving but Drowning', 1972)

Reflecting on witnessing this carnivalesque experience at sea, I have been led to consider the place of the body and immerse myself in considering the nature of well-being. The body is fundamentally a site of feeling and experience generated in place and time (Pile, 2010). While the sight of the sea can be comforting (Coleman and Kearns, 2015) and immersion can be integral to recovery and well-being (Kearns, Collins and Conradson, 2014; Foley, 2015), some people experience an abnormal and persistent fear of deep water. This experience has been labelled 'bathophobia', a condition that leads people to experience anxiety even though they rationally know they are safe from falling into, or being consumed by, oceanic depths. The word originates from the Greek *bathios*, meaning deep or depth and *phobos* meaning 'aversion, dread or fear'. According to psychologists, the ocean has always, for some, embodied extreme fears (Vanin and Helsley, 2008). Voices of the afflicted can be found floating in cyberspace.[2]

I get nervous just zooming into the ocean on Google Earth. Especially near the Mariana Trench. I have never before been afraid to do that. Now I can't even bare [sic] the thought of it.

(Raven, 16 September 2015, Fearof.net)

As a kid I would just jump in the deep and swim like a mermaid. This new fear came since my teen years. I think as we become more aware of the vastness and dangers of water and our own limitations, we acquire new fears.

(Anu, 9 September 2015, Fearof.net)

I would have a full blown panic attack every time I hit the water and start swimming because I was so afraid of the deepness and what could be in there. I would shake and shake my legs violently until I calmed down.

(Lauren, 3 January 2017, Fearof.net)

A more generalised aversion to water, or aquaphobia, is considered to be 'a marked and persistent fear that is cued by clearly discernible objects or situations' (American Psychiatric Association, 2000, 429). As in the trigger offered by my shipboard experience, there has long been knowledge of links between fear and childhood. According to Hall (1897), 'fear in some unreasonable form is almost universal at some stage of childhood' and the origins of such specific personal phobias as bathophobia can typically be linked to events in childhood and observed to intensify through adulthood (Becker et al., 2007). The therapeutic challenge is to gain and retain control over such fears and moments of anticipating the 'not-yet' (see Pain, 2000). For Aleksander (2013, n.p.), each time one is able to remain afloat in deep water even briefly, 'one adds a droplet to that well, a reserve to tap the next time you lose nerve'.

Commonly observed reactions to distress in the water are summarised by Brander et al. (2011) as including becoming overwhelmed with the situation and struggling on the spot, usually probing with one's feet and being at risk of sinking under water in indecisiveness. The depth of the fear experienced is related to how predictable the situation is and how individuals view their ability to control circumstances surrounding the phobia (Armfield, 2006).

Watery flights

a journey begins before it actually begins, and continues long after it ends.
(Symes, 2012, 66)

Since that crossing between hemispheres over half a century ago, I have lived largely at peace with the sea. I have flown over the vast Pacific Ocean more times than I can count, snorkelled off the Poor Knights Islands and been on boats in high seas off Tonga and the Chatham Islands. At times, such passages have included moments of breathing deeply and focusing thoughts elsewhere. But occasionally, in recent years, King Neptune has knocked on my door when least expected. Three instances come to mind.

First was the instance that unexpectedly launched this chapter. It was at Kawau Island, after sheltering overnight in Vivian Bay, and time for a morning swim off

Steinlager 2. Others swam out a reasonable distance; I stayed close to the yacht – treading water, enjoying the feeling of immersion but aware of the depth below. That day the shoreline seemed too far away for comfort. My legs were moving like the agitator in an old washing machine. A tipping point was reached; it was no longer enjoyable. There was no need to panic with the hull of our big red boat within reach and others within sight. But the feeling was still unsettling enough to take flight. I knew that this fear of sinking would be best arrested by making a bee-line to something more solid than the depth beneath my feet. I made my way in a hybrid breaststroke and dog-paddle to the anchor line and grasped its stretched out firmness, then touched the sleek side of the vessel. I had crossed the line. Re-berthed.

Recalling that moment of impending panic, I am taken to other times when my ambivalence with the sea has taken flight – like two summers ago, while camping up north at Mimiwhangata. My partner and I had taken out kayaks and we had paddled together offshore in a manner not unlike other summers at that bay to which we had ritually returned each January. But then it happened. I experienced a sudden sense of aloneness out there. The shore was distant and Pat was, by then, perhaps 100 metres away. Only a few years back I had paddled alone to a distant set of islands quite unperturbed. Yet on this more recent day, the frightened inner child, rendered anxious by the surrounding seas aboard the *Rangitane*, returned, calling out within me objections to the depths and all that lay beneath. Fright, flight, freeze. Propelled by a sudden and baseless fear, I paddled vigorously, but only became more aware of the depth beneath me. Finally I froze, overcome by being drawn into the depths and humiliated by my own inability to act. I raised a paddle in surrender. Joined by my companion, we paddled closer to shore, with kayaks moving as Siamese twins.

The same sensations arose more recently in the Maldives when I was accompanying colleagues to collect data on the white-sand islets in the expansive Huvadhoo Atoll. On Mahutigala, after working in the tangled vegetation identifying species and measuring profiles above sea level, we cooled off by donning snorkelling gear. Six of us paddled out through the shallow reef, navigating around architectures of coral. It had been some years since I had experienced that weightless feeling of gliding through a world owned by other creatures, seeing bright flashes of colour amid the fast-bleaching reef structures. I was happy enough to be some distance from the others, eyes behind glass mask, comforted by the sound of my breath channelled in and out of the snorkel. But then we reached the edge of the reef and a swell that was earlier merely a nudge was now pushing me about and suddenly I was out of my depth. There was nothing to stand on. Looking down and outwards, the coral formations gave way to a blue of such darkness and depth than anything could happen. Panic niggled. My finned feet came to rest on a large, smooth coral mound (Figure 14.4). I remember the exhortations at the airport terminal not to touch, take or damage coral, so I stood lightly. I was spotted by another in the team. 'Put your hand on my shoulder, and let's move back to the shallows together'. Comforting words. I think she saw that look in my eye as I retreated from the deep.

Figure 14.4 Finned feet come to rest on a coral mound, Mahutigala in the Maldives. Courtesy of Michael Hilton.

Back to shore

This chapter has provided a glimpse into the ways that my early encounters with the vastness of the ocean and exposure to the unintentionally disturbing ritual of Neptune's Court seeded an aversion to deep water. The account was triggered by a spontaneous return of memories during the onboard workshopping of chapters for this collection. I subsequently located and considered family photographs as autoethnographic materials to supplement reflection on fragments of memory. While this account of a childhood experience and subsequent 'deep encounters' are idiosyncratic, psychological literature suggests that there are widespread immersive aversions similarly felt by others. Hence, while living *with* the sea may be a sought-after and even joyful experience, immersion *in* the sea is, for me, and I am sure many others, deeply problematic.

As indicated by other contributors in this collection (see chapters by Eames, Irwin and Leather) our exposure to, with and in the sea can be a significant life experience that shapes our ongoing relationship with the sea in complex ways. A key contribution of this chapter has been its contention that sea journeys never, in a sense, end. Rather we carry with us memories and experiences – however deeply embedded in our subconscious – that mean we can still find ourselves 'at sea' despite having landed. As Symes (2012) contends, all journeys have sequels. Significantly, perhaps, my foregoing account was not a well-rehearsed story but rather a narrative that arose from the depths of my being through the dual prompts of being asked to write this chapter and that of discovering the accompanying photographs from the voyage. Even my partner of 30 years remarked in surprise, on reading an early chapter draft, that she hadn't heard of this connection between the voyage and my aversion to deep water. I could only reply that it was a connection that surfaced through the act of writing. Hence, a second contribution of the chapter is to claim a space for autoethnography that allows a 're-view' of the self through drawing on *both* memory *and* material artefacts (in this case another's photographs). This re-viewing dissolves the space between recollection and objects of representation and, as Butz and Besio (2009) claim, it allows the authors to become part of what they are studying.

In conclusion, the chapter has argued that a passion for the sea is not manifest uniformly or comprehensively. Rather, we would be wise to bear this in mind as we interact with others for whom the sea, and their experiences of it, has different meanings. By being more deeply attuned to the experiences of others we may better understand the challenges and opportunities that are available to live with the sea.

Notes

1 As with other manifestations of fear, such as that of deep water, this early witnessing generated anxiety as much centred on anticipation as on actual experience of danger).
2 These quotes are taken from comments offered by afflicted persons on the site http:// www.fearof.net/fear-of-depths-phobia-bathophobia/ (accessed 25 September 2017).

References

Aleksander, I. (2013). Facing her worst fear: An ELLE writer learns to swim at 28. *Elle*, 25 November. [Online] http://www.elle.com/beauty/health-fitness/advice/a12624/personal-essay-on-fear-of-deep-water/(Accessed 25 September 2017).

Anderson, B. and Harrison, P. (2010). The promise of non-representational theories. In B. Anderson and P. Harrison (Eds.), *Taking-Place: Non-Representational Theories and Geography*. Farnham: Ashgate, pp. 1–36.

Anderson, J. (2014). Understanding the relational sensibility of surf spaces. *Emotion, Space and Society*, 10, 27–34.

Anderson, J. and Peters, K. (Eds.). (2014). *Water Worlds: Human Geographies of the Ocean*. London: Routledge.

Anderson, K. and Smith, S. J. (2001). Emotional geographies. *Transactions, Institute British Geographers*, 26(1), 7–10.

Andrews, G. (2011). 'I had to go to the hospital and it was freaking me out': Needle phobic encounter space. *Health and Place*, 17, 875–884.

Anim-Addo, A., Hasty, W. and Peters, K. (2014). The mobilities of ships and shipped mobilities. *Mobilities*, 9, 337–349.

American Psychological Association. (APA). (2000). Specific phobias. In APA. *Diagnostic and Statistical Manual of Mental Disorders* 4th edition). Washington, DC: American Psychological Association pp. 429–450.

Armfield, J. M. (2006). Cognitive vulnerability: A model of the etiology of fear. *Clinical Psychology Review*, 26(2), 746–768.

Auden, W. H. (1950). *The Enchafèd Flood: Or, The Romantic Iconography of the Sea*. New York: Random House.

Becker, E. S., Rinck, M., Türke, V., Kause, P., Goodwin, R., Neumer, S.-P. and Margraf, J. (2007). Epidemiology of specific phobia subtypes: Findings from the Dresden Mental Health Study. *European Psychiatry*, 22, 69–74.

Borovnik, M. (2004). Are seafarers migrants? Situating seafarers in the framework of mobility and transnationalism. *New Zealand Geographer*, 60, 36–43.

Brander, R. W., Bradstreet, A., Sherker, S. and MacMahan, J. (2011). Responses of swimmers caught in rip currents: Perspectives on mitigating the global rip current hazard. *International Journal of Aquatic Research and Education*, 5(4), 476–482.

Brown, M. and Humberstone, B. (Eds.) (2015). *Seascapes: Shaped by the Sea*. Farnham, Surrey: Ashgate.

Butz, D. and Besio, K. (2009). Autoethnography. *Geography Compass*, 3, 1660–1674.

Chang, H. (2007). Autoethnography: Raising cultural consciousness of self and others. In G. Walford (Ed..), *Methodological Developments in Ethnography (Studies in Educational Ethnography, Volume 12)*. London: Emerald Group Publishing Limited, pp. 207–221.

Coleman, T. M. and Kearns, R. A. (2015). The role of blue spaces in experiencing place, aging and wellbeing: Insights from Waiheke Island, New Zealand. *Health and Place*, 35, 206–217.

Connell, R. W. (2005). *Masculinities*. Berkeley: University of California Press.

Davidson, J. and Milligan, C. (2004). Embodying emotion sensing space: Introducing emotional geographies. *Social and Cultural Geography*, 5, 523–532.

Foley, R. (2015). Swimming in Ireland: Immersions in therapeutic blue space. *Health and Place*, 35, 218–225.

Hall, S. G. (1897). A study of fears. *The American Journal of Psychology*, 8(2), 147–249.

Hasty, W. and Peters, K. (2012). The ship in geography and the geographies of ships. *Geography Compass*, 6, 660–676.

Honoré, C. (2004). *In Praise of Slow: How a Worldwide Movement is Challenging the Cult of Speed*. London: Orion.

Kearns, R. (2015). Understanding the heart of place. In S. C. Aitken and G. Valentine (Eds.), *Approaches to Human Geography: Philosophies, Theories, People and Practices* (2nd Edition). London: Sage, pp. 239–246.

Kearns, R. A., Collins, D. and Conradson, D. (2014). Healthy island blue space: From space of detention to site of sanctuary. *Health and Place*, 30, 107–115.

McKay, S. C. (2007). Filipino sea men: Constructing masculinities in an ethnic labour niche. *Journal of Ethnic and Migration Studies*, 33, 617–633.

McGlashan, D. (1994). *Anchor me (song)*. *Salty* Auckland, NZ: The Muttonbirds, EMI.

Merchant, S. (2011). Negotiating underwater space: The sensorium, the body and the practice of scuba-diving. *Tourist Studies*, 11, 215–234.

Muncey, T. (2005). Doing autoethnography. *International Journal of Qualitative Methods*, 4, 69–86.

Officer, A and Kearns R A. (2017). Experiencing everyday train travel in Auckland: An autoethnographic account. *New Zealand Geographer*, 73(1), 97–108.

Olive, R., McCuaig, L. and Phillips, M. G. (2015). Women's recreational surfing: A patronising experience. *Sport, Education and Society*, 20, 258–276.

Pain, R. H. (1997). Social geographies of women's fear of crime. *Transactions of the Institute of British Geographers*, 22, 231–244.

Pain, R. H. (2000). *Place, Social Relations and the Fear of Crime: A Review Progress in Human Geography*, 24, 365–387.

Pain, R. H. and Smith, S. J. (2008). *Fear: Critical Geopolitics and Everyday Life*. London: Routledge.

Peters, K. and Brown, M. (2017). Writing *with* the sea: Reflections on in/experienced encounters with ocean space. *Cultural Geographies*, 24(4), 617–624.

Phoenix, A. and Frosh, S. (2001). Positioned by 'hegemonic' masculinities: A study of London boys' narratives of identity. *Australian Psychologist*, 36, 27–35.

Pile, S. (2010). Emotions and affect in recent human geography. *Transactions of the Institute of British Geography*, 35, 5–20.

Poe, E. A. (2017). *The Complete Works of Edgar Allan Poe*. Musaicum Books. [Online e-book].

Richardson, K. P. (1977). Polliwogs and shellbacks: An analysis of the equator crossing ritual. *Western Folklore*, 36(2), 154–159.

Schellhorn, M. and Perkins, H. C. (2004). The stuff of which dreams are made: Representations of the South Sea in German-language tourist brochures. *Current Issues in Tourism*, 7, 95–133.

Smith, S. (1972). *Collected Poems of Stevie Smith*. New York: New Directions Publishing Corporation.

Spence, E. (2014). Towards a more-than-sea geography: Exploring the relational geographies of superrich mobility between sea, superyacht and shore in the Cote d'Azur. *Area*, 46, 203–209.

Stocker, L. and Kennedy, D. (2011). Artistic representations of the sea and coast: Implications for sustainability. *Landscapes: The Journal of the International Centre for Landscape and Language*, 4 (2)[Online] http://ro.ecu.edu.au/landscapes/vol4/iss2/28 (Accessed 25 September 2017).

Symes, C. (2012). All at sea: An auto-ethnography of a slowed community, on a container ship. *Annals of Leisure Research*, 15(1), 55–68.

Tolich, M. (2010). A critique of current practice: Ten foundational guidelines for autoethnographers. *Qualitative Health Research*, 20, 1599–1610.

Tolich, M. (2012). My eye-opening midnight swim: An Outward Bound autoethnography. *New Zealand Journal of Outdoor Education: Ko Tane Mahuta Pupuke*, 3(1), 9–23.

Vanin, J. R. and Helsley, J. D. (2008). *Anxiety Disorders: A Pocket Guide for Primary Care*. New York: Humana Press.

Vannini, P. (2013). Slowness and deceleration. In P. Adey, D. Bissell, K. Hannam, P. Merriman and M. Sheller (Eds.), *The Routledge Handbook of Mobilities*. London: Rouledge, 116–124.

Wall, S. (2008). Easier said than done: Writing an autoethnography. *International Journal of Qualitative Methods*, 7, 38–53.

Weaver, A. (2005). Spaces of containment and revenue capture: 'Super-sized' cruise ships as mobile tourism enclaves. *Tourism Geographies*, 7(2), 165–184.

15 Conclusions

Learning to live with the sea together: opening dialogue, creating conversation

Kimberley Peters, Alistair Moore and Mike Brown

In the introduction we set the scene for the unique conference from which this book developed (see Chapter 1). Organised by Mike – maritime scholar, outdoor educator, sailor – the conference had an unconventional, yet appropriate setting – an 80-foot former race yacht. 'Seascapes: Living with the Sea' would be held at sea over three nights and four days in the Hauraki Gulf, New Zealand. It would be based on the *Steinlager 2*, a famous boat operated by the New Zealand Sailing Trust. The event would bring together scholars from cross-disciplinary perspectives (education, law, architecture, planning, geography) and from stakeholder groups and activist organisations. It would bring them together to sail. To talk. To write. To learn. And ultimately to ask, 'What is it to live with the sea?'

In this final chapter we reflect on the conference from several perspectives – ours as editors, together with the skipper of the vessel, Alistair Moore.[1] As a commercial vessel *Steinlager 2* is required to comply with New Zealand Maritime regulations. These requirements relate to the structural integrity of the vessel, the equipment to be carried and the number, and appropriate certification, of professional crew. As an ex-Round the World Race boat *Steinlager 2* is both simple, in regard to the standard of accommodation, but equally complex in regard to its sail handling systems. The loads imposed on ropes can be huge and careful oversight is required to ensure that participants are appropriately supervised and kept clear of any potential hazards. On our trip the skipper, Alistair (or Al, as we came to know him), was ably assisted by two mates/deckhands. Over the four days both of us had the opportunity to chat to Al and it became clear that being a skipper was more than a job. It was interesting to see how Al would join our discussion sessions, and whilst he sat on the periphery and kept his own counsel, it was clear that some of the topics presented resonated strongly with him.

Al's role as skipper clearly involved making appropriate nautical decisions based on the wind direction, tides and the medium-term weather forecast; but it also involved more nuanced and 'people-centred' skills. For this voyage, or in fact any voyage to be successful, the skipper needs to be sensitive to the ability/desires of the crew, be adaptable in order to respond to their level of comfort – yet provide an appropriate sense of challenge and excitement – and to create an effective learning environment. Clearly one of the elements of this successful conference was Al's responsiveness to our needs, to ensure

Figure 15.1 Alistair Moore, standing centre. Opening day of the conference. Photo by Mike Brown.

that we had adequate time to engage in academic pursuits but also to provide enjoyable and safe sailing experiences.

For Kim, this was crucial. As we have reflected elsewhere,[2] the conference presented a first-time sailing experience and direct engagement with the sea for some, and unexpected encounters for long-time, established sailors. For Kim, who had never sailed before, there was a sense of unfamiliarity, anxiousness, sickness at living with the sea for four days. For Mike, an experienced sailor, there was a realisation that living with the seas could come in many forms. For those who chose not to swim, to stay within the safety of the vessel, or to watch from the shore, the sea still touched them in no less significant ways. The experience of learning together, of opening dialogues, or creating conversations about the sea, *at sea*, was profound.

Yet, following the conference and our own reflections we kept returning to the 'quiet role' played by 'quiet al', our skipper. How, for example, did he come to skipper a sail training vessel? We wondered not only about his career pathway but also his personal connection to the sea. As academics we came to the conference with various analytic or theoretical perspectives informing our work. What was Al's take on what he was doing? We sensed, in our chats with him, that there something more than the enjoyment of sailing that took him away from his wife for four or five days at a time. Why give up the comforts of home to share meals with up to 20 strangers every week? We decided that we'd like to hear Al's story, to add what could be a different perspective to our understandings of living with the seas. What was it like to live with the sea when you engaged with it daily? We sought to understand a professional mariner's understanding of *Living with the Sea*.

The edited transcript below allows Al's voice to stand alongside those of the other contributors of this edited collection. In reading the transcript back it became apparent that the distance between the academy and professional practice is not as wide as it can seem. We often speak a common language about the sea, even though how we live with it may be different. As the metaphor at the end of this piece confirms, when it comes to living with the seas, the most significant point of all is that we're in the same boat, together.

***

MIKE: *So I guess a good place to start would be what got you into sailing?*

AL: You see sailing was something the family did together, it was something that we just did. Where we lived, mucking around at the bottom of the garden, with the tide coming in, as little kids. Most weekends and most summers were spent on Dad's cruising boat. So it was, I didn't feel that it was anything different or special. It was just that you know you were around it all the time, so . . . it was . . . became second nature you know, by five years old I was using an outboard on a dinghy and you know stuff that most kids, most parents wouldn't let their children do, but it was just something you'd been around since you were two weeks old . . . the water and messing about in boats became second nature.

MIKE: *. . . your Mum was into it as well?*

AL: Yeah, though I think she suffered it to be with Dad, you know. She certainly enjoyed the good weather . . . but suffered the bashing back from the Bottom End [*eds.: area on the eastern end of Waiheke Island*] blowing hard from the south-west. Whereas Dad really got a kick out of that you know. Man versus the elements; and I think I got a bit of that off him. At about, ah, just before my sixth birthday, I got my first sailing dinghy. I got an Optimist and started racing that at the local club. Yeah, and that was when you know, you started to realise how much you could learn. You thought you, you thought you had it all under control, and you didn't. And learning that as a young boy, you know, going from 'I've got this, I've got this' to, 'shit, I really don't have this, what do I do now to get myself out of this situation?' All sort of big lessons learnt before ten years old.

MIKE: *So do you think you learnt those lessons at the time, or is it something you think you now reflect on?*

AL: I think I probably reflect back on, you know. You know I remember being amazingly confident with just my first couple of races and then . . . realising as time went on that I didn't have all the answers. We grew up, up the Waitemata [*eds.: the main harbour on the east coast of Auckland*] on an island, you know, with a little road that went up to it. Herald Island,

and so we were surrounded by water. I had a couple of liberal teachers that suggested that I was to go sailing instead of doing homework. You know and it was escapism a little bit but it was also learning to be reliant on yourself. You got yourself into a situation, it was up to you and no one else to get yourself out of that situation. You know, exploring up the mangroves as a kid, the wind would run out and you having to row home or whatever, those kinds of things as a boy. And ... through the racing side of it, I certainly developed a competitive attitude, you know.

KIM: *So when did you make a kind of transition between just going out on your boat, you know just part of your daily life, to you starting to race in a sort of competitive way?*

AL: I guess as a natural progression of being surrounded by a couple of mates who were competitive and you know, and then it became any time we were on the water together, it became a race, right?

KIM: *Yeah.*

AL: Yeah, that all started around Herald Island between seven and eight and progressed through. I changed clubs because I didn't feel the club I was at was competitive enough. At 11 I moved to another club and ... that really shocked me, that was a big step up in competitive sailing. I had a very poor season my first season there, a pretty good season my second season and then my third season, I was a big kid so I was in a Laser [eds.: *popular sailing dinghy*], so there were kids that were a lot bigger, a lot older than me. That was the steepest learning curve I ever went through. And started to get a little bit frustrated with it, at 14 was, you know ... I'm not getting the results I feel I deserve. And it was my father that suggested that, well you know, get out of dinghies, get into bigger boats, because of your size, you know. And yeah ... before my fifteenth birthday I had a 24ft yacht that my best mate and I had bought together. That was the beginning of, I guess, passage sailing for me, independently. I went down as far south as Hot Water Beach and Tauranga and then you know as far north as ... as Leigh and Omaha and up those areas, as a 15-, 16-year-old. Off the back of that, I managed to secure a position in the youth development programme with the Royal New Zealand Yacht Squadron, so I did two years under Harold Bennett, who was a very, very good coach. I think that, that very intensive ... amount of sailing and of course the level of the coaching that came along with it, was what took me from a weekend sailor, you know, someone that was competent and safe, to somebody that was ... far more aware of the subtleties, you know, far more complete. But it was a massive commitment, it was, you know, both days of every weekend and two nights a week for two years. It was a ...

KIM: *And were there ever any times when you sort of thought, I don't want to do this?*

AL: Mm ... not really, no. There was something about the camaraderie of the group of 30 young people that were there. The boats I really loved, they were very, very fast, quite a difficult boat to sail well but it gave you an amazing reward when you did everything right. Yeah, and there was, it was just starting to be a pathway for youth to, to make it a career, rather than just a pastime, and this is early nineties, so yeah, '92, '93.

KIM: *Because I guess that's it for me, not being someone who is from the sea or spent a lot of time at sea or anything like that, it's like 'what is that thing that keeps drawing you back'? You know, as a young person who was brought up with the sea – what is it that keeps you there? What is it that tethers you to that space? You're not bored of it ...*

AL: No, I think it is ... it is because of the subject matter itself, you know ... just being on the water is ... this is normal, you know, the boat is meant to lean over, that kind of stuff. From a very young age, I was comfortable with [it]. The thing that made me, makes me stick with it is that it is so very varied in its similarities. It's, when I go cruising, I ... I want to be sailing so I can have the motor off so it's quiet. I'm in harmony with my environment, and I'm, I am ... not ... not having any detrimental effect on it, whilst enjoying it, whilst travelling through it, whilst being amongst it you know. I'm not putting in undue amounts of energy to get the boat moving. I enjoy the journey, but then there's destination. What's over the next horizon, what's round that next bend? Ah it's tremendously rewarding, the exploration side of it especially, you know, what is here? What are we going to find?

MIKE: *So what was running parallel, so when you were doing the squadron, did you stay at school ...?*

AL: Yeah, I was in my last year at school and first year at university. So ... with the racing side of things, the two, the cruising side was very separate, you know. It was something I did with mates that weren't sailors, and it was ... taking a lot of responsibility on your shoulders, because you're bringing people with you that are outside their comfort zone, versus the racing side of things, where everybody's at a very competent level to begin with and it's about ... using the environment and your skillset to try and get one up on the next person, you know, it's a little bit of chess, or a lot of chess.

KIM: *Yeah.*

AL: So you know the cruising is a wholesome, family ... outlet for, for gaining new experiences, for gaining a new horizon, a new ... scene. The racing side is using the environment and all these, you know, experience gives you that local knowledge or that subtlety that you know that as the sun clocks that way, the wind's going to do this with it, on this kind of day, you know? And then putting that into your game plan to try and get one up on someone. So it ... the two of them dovetail very well but, ah, do stand alone in my mind, you know, the difference between racing and cruising.

KIM: *So yeah, I mean that is the thing that I kind of learnt on the boat. You guys do so many different things, all at once, like all those things you're just describing about, you're thinking about all these different practices. You're thinking about the boat, you're thinking about the weather, you're thinking about how all of these things ...*

AL: The mechanics, yeah.

KIM: *... you know ... But I was quite interested in what you were saying there about ... you know for you, this is just, it is just, it's second nature I think was what you said.*

AL: Mm.

KIM: *So what's it like working with people, people like me, who don't know what to do ... is it odd to kind of see people for whom this is so unfamiliar?*

AL: No, no, I don't think it's odd. I think it, I mean it's ... it makes me realise how lucky I am, it also ... I'm lucky to be able to give people this first experience, because I know, I have heard of and have seen, people have terrible first experiences on boats because they don't have the right kind of mind-set around them. The people that are either barking orders or not giving enough information or ... and it gets people really on edge. So I ... I ... think I have a skill at making people feel at ease, you know, letting them know when's a good time to panic, that kind of stuff. So no, I'm, I'm very lucky and ... privileged to be in the position I am, to give people these experiences. Yeah. I have a reasonably, not unique, but a special upbringing that has afforded me a slightly different skillset to someone that say, well someone that didn't spend half their life at sea.

MIKE: *So what was your trajectory, so you did, did you do any course after doing outdoor education at uni?*

AL: Within a couple of months from graduating, I'd worked out that I wanted to earn my money at sea, rather than use it as my recreation time. A career at sea was what I wanted, so I got a job on a game fishing boat, running out of the Bay of Islands, as a deckhand, to get commercial sea time, to then go back to school and do my commercial tickets. So I did nine months at sea on the fishing boat, which I hated, but afforded me good quality sea time, and helped me through those exams. Pretty much straight out of graduating with a commercial licence, Louis Vuitton [*eds.: major sponsor of the America's Cup*], through a family friend, got in touch and asked me if I would come and help them run the media boats for the first America's Cup down here. So I had five big rigid hull inflatable boats that I had to put into survey. I did all that kind of paperworky side of things, and then find crews for them. And during that Cup, Lady Blake, Peter's wife, would come out on the boat that I ran and we got on quite well. Towards the end of the America's Cup, I gave her my CV and asked cheekily if she would put that in front of her husband. I'd just been offered a job in France and I was hoping that I could get some advice off him. He called me up that night and said, come and see me in the office. He told me the job offer was a solid one, and it was around a racing boat in France, for a well-known sailor, and Pete said, 'yes, fantastic guy, you'll learn French, you'll learn good seamanship, but here's another option', and he pointed out the window and said, you can start on Monday ...

MIKE: *On what?*

AL: On *Seamaster*, which was then Cousteau's Antarctic Explorer, and to be offered that from your childhood hero ... yes! ... no you don't have to pay me. So I did seven months refit work on the boat here, and then a trip around

New Zealand testing it, and then off to South America and down to Antarctica and ... And when we got back from Antarctica, that's when he said, yeah Alistair, you're actually pretty handy, um would you skipper a boat up to the Buenos Aires? So we got there and did a tropical re-fit in Buenos Aires. Getting it ready for hot climates. Peter sailed with us up to Rio, and then he got off and flew to Germany and he then joined the boat a little later. We went up the Amazon and up the Rio Negro, and then I got onto the small boats with three other crew members, and we continued on, and *Seamaster* went back down the Amazon with a plan of coming up the Atlantic and up the Orinoco and meeting us coming down there. Peter got shot and no one else was insured to skipper the boat, so they flew me out of the jungle, got me back to the boat, and I skippered it up to Trinidad and Tobago. I then came back to New Zealand ... and just through being here at that time, I fell in with a couple of guys that I used to race with and got a spot on the crew with the Danish match racing team. So I did that next year, 2002, racing for living, all over the world. We were one of the few teams that lived off winnings, most teams were tied to an America's Cup team. We were ... four Danish university students and myself. And ... yeah, did the, did the Swedish Match Tour, which was the world matching racing tour events, and ended up winning, ended up being the world champions at the end of that year. And after that season finished, I started working for a Frenchman, with his cruising boat and his family, just helping him out, and then ended up brokering the deal where he bought *Seamaster* off Lady Blake. I then ended up running her for another three and a half years for him, to Greenland and Iceland, had a big re-fit in France and then down, back down to Antarctica via Argentina, the Falklands and South Georgia. And after that I came home and started running the America's Cup tourist boats down here, as a fill-in job, and so ... yeah. Did that for a year, and then ran *Lion New Zealand* up and down the coast for a year, as a tourist, kind of back-packer operation, as it were, and took *Lion* up to Tonga, ran her as a whale-watching boat. And then I got involved with the NZ Sailing Trust. *Lion* was gifted to the Trust in late 2008 and I worked, for free, for about six months, trying to get it up and running. I then got offered a job in Europe, so took off up there, worked for a couple of different race teams in Europe. Then came home, did six months here and then got offered another job up there, doing youth development. Then I worked for Oman Sail for a couple of years, teaching young Arabs how to become offshore sailors. That was interesting, hard work, based out of Oman, so sailing in the Persian Gulf, surrounded by oil tankers and drug smugglers. I lived on the South Coast of England for a couple of years, in Cowes and the Isle of Wight. Fantastic place ...

MIKE: *what did you do sailing-wise in the UK?*

AL: Working for Ellen McArthur's Offshore Challenges, so preparing an Open 60 for the Vendee Globe [*eds.: single-handed non-stop round-the-world race*]. Then Extreme 40s, doing a lot with the Extreme 40s circuit. And then ended up working for a team called Pindar, yeah, nine months with them as the team engineer, getting the boat ready for the Vendee, the next one. But when the, the

Trust acquired *Steinlager*, I was halfway around the world with a Volvo 70. And they got in touch and said, we've got more clout now with the two boats, there's a full-time position here for you if you'd like it. It was just the timing sort of, it fitted with, you know, coming home and SJ and I being in one place that we both wanted to be in, rather than me in the Middle East and her in London, and every couple of months we'd get a weekend together.

KIM: *I'm just listening and all I can think is ... that's a lot of movement, you know, and what becomes, is there ... did you need a point of stability? Was there a point of stability in all of that? What was ...*

AL: So I guess I had mum and dad and their home, you know, that was always in the back of the mind that if it all turned to custard, I could go and live with mum and dad, which was a bit depressing but I ... I was really fortunate to have strong parents that told me ... suggested that I do what I love, rather than chase the almighty dollar. Which is an unusual thing for baby boomer parents to say. But yeah, that came from a father who plumbed his entire life and his spare time was spent sailing, and he thought it strange to be doing something you don't like to have some spare time to do something you enjoy – do it the other way round. Yeah, and so through their support, I always sort of backed myself. This current gig is going very well, but it will come to an end. They all come to an end, then there'll be something else you know.

MIKE: *So what is it about the Sailing Trust that, that you're here? I mean obviously there's other things going on, you know, family ties ... But I mean there's other options for you, I guess for you in the maritime industry here.*

AL: Yes, there, definitely there would be. There's something about the ability to ... to not have, not have a plan. With the Trust and the trips that the boat does, it's unlike any other ... vessel. With other boats it's normally, we'll be here at this time, and we'll be here at this time. Here, having the flexibility to deliver an experience that fits the group ... that's a challenge the whole time. You know, two trips are never the same, because you're always trying to deliver it for the group, rather than yourself. I get a real kick out of seeing either the really cocky people come down a level or two, and the really timid people come up a level or two, and I think ... time living on a boat is ... it's compressed space, but I think it also compresses time. I think you can get, you get to know people far quicker, far better, because ... they're outside their comfort zone and in a very level hierarchy. Everyone's dealing with the same conditions, there's no, very little differentiation between crew and, and guest, you know it's ... it's a great way of knocking any kind of social barriers down and getting to the nitty gritty of what gets a result and what doesn't. So yeah, I get a kick out of it, a real kick out of it.

MIKE: *So what do you think are the outcomes for the students?*

AL: I think the biggest thing for the young ... well not even ... you know ... say up to 19-year-olds, there is a sense of ... them finding some inner quality in themselves that, that proves to them that they can. It doesn't matter what it is, but they, you know, through tenacity, through focus, through a little bit of energy put in, massive rewards can come out of it. And something as simple as steering the boat when the, when the person that's steering the boat realises that ... maybe they're the only one actually concentrating and everybody else is having a chat, that they're responsible, their actions are keeping the rest of the team happy, safe and travelling in the right direction. Yeah I think ... it's that whole ... it compresses it so much ... oh compress is the wrong word, but there's so much to experience, for someone that doesn't find it, you know, hasn't had the, that experience before, it lowers the defences and ... there's a, there's a ... most of them become sponges, soaking the stuff up.

KIM: *I find that really interesting, what you'd just said. You know ... people have to ... everyone's in it, it's a leveller, you know, because you have to have that responsibility. People have different roles at different times and it just reminds me of what you were saying earlier, about being a kid and being on a small boat and ... quickly learning that something could go wrong and you've got to ... and it's almost like you're seeing other people have those same experiences ...*

AL: Oh I'd agree, yeah, that's ... yeah, I mean those three or four 'holy shit experiences' you have when you're a kid when you've got it wrong. You don't want, I mean we can't facilitate those kind of ... because it's too dangerous, but it's, if you're getting a little bit of that going though, in a safe ... and managed environment, yeah, you, it ... the amount you grow from those experiences is, it's very, it's actually difficult to, to quantify, and that's something that we're struggling with in the Trust, is the ability to actually find, you know, a way of quantifying the experience into, is it successful or is it a failure? Have we achieved the goal, have we not?

KIM: *Yeah, because how do you measure, how do you measure that kind of ... I think it's a really interesting question because I've been trying to write with Mike for the last couple of weeks about how do you, how do you come to realise what the significance of any given experience is? It's actually really difficult to try and put some sort of measure on it, that then allows you to ... And I can imagine when you're involved in a Trust and you need to know, you know are there things you're doing, are they, you know working ...?*

AL: Yeah.

KIM: *Do you think ... have you had any sort of experiences of what they seem to take from like the sea itself, or any sort of sense of ...*

AL: Of vastness

KIM: *Mm.*

AL: When we talk about you know, we'll say ... we'll do a big sail out to Great Barrier, take six hours and as you're getting closer to the anchorage, you can see them starting to get a little bit sort of white line feverish. There's that little bit of ... anxiousness. And then you just like, you know, you mention, well if you just want to hang on another four days, we'll be in Fiji. And that sort of suddenly goes ... Fiji, we can go to Fiji? Yes, you know, if you can keep the boat safe, if you have enough provisions on board, this boat is more than capable of getting us anywhere on the planet ... if you've got the time. And they, that is when they start thinking, you know, they've done this six-hour sail, they've done 60 miles, so only another 940 to Fiji. And it's that ... being connected ... to the whole world.

KIM: *Yeah. Because I guess that's the thing, it is all about sort of perspective isn't it and you know, you talk about the sea just being that space of second nature and you know that's how I feel being in a city, you know, cities make me feel comfortable. The sea makes me feel uncomfortable, and . . . it's amazing how those experiences you can have when you're younger then are so formative and sort of carry with you and . . .*

AL: I think when we have a, when you have a trip that has inclement weather and . . . your physicality is tested, you know, because you're just hanging on . . . the impact is far higher than on a trip where it's benign weather, and it's flat and easy. It's . . . it's good to have them a little bit scared for a little bit of the trip. You get a far better result at the end of it.

KIM: *There was a little bit of me that was scared on our trip! What was it?! That second, what day was it, that, the second day we sailed, the day when it was quite . . .*

AL: It was quite windy.

KIM: *I wasn't worried. It was odd, it was odd, it's an odd sensation isn't it because it's that, you know you're in the hands of an expert who knows what they're doing, and as Mike kept telling me, 'you're absolutely safe'. But at the same time, there is that bit of you that is quite humbled by the environment that you're in, and respectful of that environment. I guess what makes me so intrigued by it all is how the sea has a vastness to it. I think that's part of it, the vastness of it, maybe the fact it's always moving, maybe the fact it's . . .*

AL: Because it's, well it's fluid, it's fluid, it is . . . you know . . . what could be an absolute cake walk one day, the next can smash you round and take nine times as long, to do the same passage. So there's that volatility, or unknown-ness, you know. It's this thing that is always in flux, and yet it is the only horizon that hasn't changed in 10,000 years, you know, it's still that sea horizon. But it's always changing.

KIM: *I do remember on the boat you saying about the horizon being this one constant in all this world of shift . . . We were talking about Dave's paper [eds.: see Chapter 11] about thinking about the oceans and sustainability and you know the kind of . . . the sort of . . . I don't know, the vulnerability of the environment, and what we're, what we're doing to the seas and how every-thing is changing. And then . . . this notion that so many things are changing, yet there is, there's this horizon . . . this thing that we can't actually change, you know, there's nothing, there's nothing we can do to change the horizon . . .*

AL: Mm.

KIM: *. . . you know this constant . . . which is quite a, I think, quite a powerful notion . . . there's something we can't touch, we can't . . .*

AL: When you, when you're outside of land, you know, when you're . . . there's something really reassuring about the horizon being there.

MIKE: *But what I think is interesting is what's going on beneath the horizon, so you know we know that there's nowhere now on the planet you cannot find microscopic particles of plastic in marine life. The sea in some ways acts as a camouflage doesn't it, because it's green or it's blue and it . . .*

AL: Hides anything that we dump in there.

MIKE: *Yeah. Till it comes to a tipping point I guess.*

AL: Mm, exactly.

KIM: *Do you think that through the sort of things you do ... the kids that come on [board] ... do you think that they gain a sense of any sort of environmental kind of ethic ... does that come into it?*

AL: For me, personally yes, I mean ... each person is different and each person takes ... chooses their level of responsibility as custodian of the planet. But I think if you can ... I mean the connectivity that the ocean gives the world, if you can lean on that and get people to realise just how interdependent each little micro ... community, environment is reliant on the next one and the next one and the next one and ... you know, how sharks are good for fish stocks because they take off the weak. The whole interdependency of, of everything, in the marine environment, is a great lesson to be taught and seen first-hand. So if you can see the gannets feeding with the dolphins or ... or the terns with the Kawai, you know, that kind of, those kind of relationships ... a bird can't speak fish, fish can't speak bird, but there they are you know ... hanging out together, and benefiting off each other's activities. You know? And then you go and show the fishing trawler going through the middle of it, or a beach clean-up, where we'll find you know 100 cigarette lighters, that kind of stuff.

MIKE: *So you do a bit of beach clean-up?*

AL: Ah, definitely yeah, and catalogue what we've got and where we got it from, and it goes through Sea Cleaners, I've forgotten the name ... a good charitable trust that do the big clean-up round here, so a little bit with Sustainable Coastlines.

MIKE: *OK, so there's an equal, quite a long, strong ecological focus on ...?*

AL: Strong, yeah. Well we can offer, and it's about offering to the group, you know, there's definitely a base level of environmental education as well. But we can certainly ramp it up and make it all about that.

MIKE: *So what would you do if you weren't doing this job?*

AL: I hate to think ... if I didn't have this job ... I don't, yeah, no, it would be something in the education field, something around sailing, yeah.

MIKE: *So clearly you started off life as a sailor first. What you're doing now is education, so did you see, was there some point there was a sort of a transition?*

AL: No ... I've always, through high school life, I was always coaching junior teams, so I think I've got a natural leaning towards education. When you're teaching something you're passionate about, I think it makes it very easy. Peter Blake was one of my biggest heroes and to be able to ... use him as a ... benchmark, role model, and his lessons and his life be used as a case study for what one person can achieve, and they can mobilise an entire nation behind them, I think that's a ... you know valuable thing to celebrate. Yeah, because if you can get just one of, you know, since the Trust, since I've been back with the Trust in 2000 and ... late 2012, we've had over 4,500 kids through the boats. Now if just one of those can have a life half as a successful as Peter Blake ... I enjoy what I do, definitely, I'm sure you do too.

KIM: *It makes me want to learn more, you know I don't know if it's an academic thing that you constantly want to learn, you know, academics are always kind of intrigued, want to learn things, want to know about things, but it definitely makes you want to learn more. I guess for me, I've always been really intrigued by the fact that there's two things going on here, you know, you've got the sea and what the sea is, and then you've got the boat that enables you to access the sea, but you also have to be able to master that particular technology and the skill that goes with that. You know I just, I was on that boat and I just thought, you know, I don't even know where you, I don't even know where you'd start, I don't even know how you'd begin to . . . just so much of it looks so complex and so . . .*

AL: But that's, when you start looking at things, you know, you break it down and go down to the simplest of principles. What is the wind direction? Start with the really simple stuff. Where am I trying to get to? What is the relative wind direction to where I'm trying to go, you know and . . .

KIM: *I wonder how many . . . how many people who come in to the boat, how many of them maybe go on and do anything to do with the sea, or how many of them sail again . . . do you get any sort of sense of that really?*

AL: Yeah, well I mean we'd probably look at, at an average . . . let's say an average group, two or three would have sailed before. And you'll find that they are reinforced to go and do it again, massively. And you'll get two or three complete novices, on an average, in each group, depending on social . . . socio-economic things. So yes, six people on each trip, if you've got an average of 22 on board, will actively get involved in sailing. But there's, I think the measure needs to be more around social value, worth, that the person puts on themselves, you know that's, it's not about, it's wonderful if they fall in love with sailing and the sea, but that's secondary for me, it's more about learning about what makes them happy, about what makes them sad, puts . . . puts a value on life outside of money. A value on life that is more to do with looking after this, this group, you know, it's such a great analogy for the planet. This is our boat, this is what keeps us alive, these are the people on it that we've got to look after, you know, the boat comes first, if we look after the boat, all the people are safe. And you can take all those lessons and go, 'OK, human race, the earth's your boat, deal with it, you know'. And it, you can get that kind of thought process to go from, you know, respecting one another, picking up after yourself, doing your fair share on the boat, and put that across to being part of the human race, Jesus Christ, that's powerful.

KIM: *And, then it goes back to that word you used earlier, which is 'compressed', right, which kind of stuck in my mind, and I just keep wondering. That's why this is so interesting to me, because I just think when you're trying to teach that and you're doing it in a classroom or you're doing it on land, when you can just pop off to a shop afterwards and you get your can of Coke or . . . you can't do that on a boat, once you're there, you're there!*

AL: Mmm

KIM: *And you can't ... And that was one of the things that struck me is that we're ... there were times on the boat when I sort of thought, ooh, I'd quite like five minutes to myself ...*

AL: Myself now please, yeah.

KIM: *And it's kind of, and you can't, and you're in that ... that compression of space ... even though you're in this open sea, or even if you're, you know, even if there's an island or what ... you're ... there is ...*

AL: There is wild environment, you're still tied to the boat.

KIM: *Yeah.*

AL: Because the boat keeps you safe.

KIM: *Mm.*

AL: So this notion, this notion of I'm just going to have five minutes to myself, I'm just going to nip down to the dairy, is false, because you're not removing yourself from the boat.

KIM: *Yeah ...*

AL: You know?

KIM: *Well you can't ever remove yourself from the boat, can you?*

AL: Well not unless you get in a bloody space ship can you?

Laughter all round.

Notes

1 As this book goes to press, Al and his wife Sarah Jane are busy refitting their boat, to head offshore on their own cruising adventures. This has required that Al leave the employment of the New Zealand Sailing Trust. We wish both Al and SJ all the best on the next set of voyages in their lives.

2 Peters, K. and Brown, M. (2017). Writing with the sea: Reflections on in/experienced encounters with ocean space. *Cultural Geographies*, 24(4), 617–624.

Index

Milton Keynes UK
Ingram Content Group UK Ltd.
UKHW040106071024
449327UK00019B/850

9 780367 586928